# BIOMEDICAL PROBLEMS ON BOVINE SOMATOTROPIN USE IN MILK PRODUCTION

# BIOMEDICAL PROBLEMS ON BOVINE SOMATOTROPIN USE IN MILK PRODUCTION

**Gianni BENZI**
*Institute of Pharmacology*
*University of Pavia (Italy)*

**British Library Cataloguing in Publication Data**

Biomedical Problems on Bovine Somatotropine
Use in Milk Production
BST, Health, Metabolism, Dairy cattle,
Dairy Cows, Milk
1. Benzi Gianni
I. Title

ISBN 0-86196-2923

**Editions John Libbey Eurotext**
6, rue Blanche, 92120 Montrouge, France.
Tel.: (1)47.35.85.52

**John Libbey & Company Ltd**
13, Smith Yard, Summerley Street, London SW18 4HR, England
Tel.: (01)947.27.77

**John Libbey CIC**
Via L. Spallanzani, 11,
00161 Rome, Italy.
Tel.: (06)862.289

© 1990, Paris

Il est interdit de reproduire intégralement ou partiellement le présent ouvrage – loi du 11 mars 1957 – sans autorisation de l'éditeur ou du Centre Français du Copyright, 6 *bis*, rue Gabriel Laumain, 75010 Paris.

# Contents

**1. General biological characteristics of bovine somatotropin (BST)** ............................................................. 1

    Recombinant and pituitary-derived somatotropins in a complex interrelated neuroendocrine hierarchy .................. 1
    Influence of BST on the neuroendocrine system ............... 3
    Indirect influence of BST in causing the increased milk yield ............................................................................. 5
    Biological modifications by BST treatment ........................ 8
    Mode of action of BST in milk production ...................... 9
    General remarks on the biological action of BST ............ 11

**2. Influence of BST on the metabolism of the whole animal and of specific organs** ...................................................... 13

    The activity of BST on carbohydrate and lipid metabolism in the whole animal ....................................................... 14
    BST interaction with the endocrine system ....................... 15
    Eumetabolic (homeorhetic) effect of BST ........................... 16
    Effects of BST in lactating animals ..................................... 17
    General remarks on the metabolic effect induced by BST ....... 18

**3. Long-term effects of administration of BST in dairy cattle** .. 19

    BST long-term effects on milk production and composition ............................................................................... 19
    Voluntary dry matter intake and calculated energy balance in animals during normal lactation and when BST treated .................................................................... 22
    Relationships between response to the BST treatment and management conditions ............................................... 23
    General remarks on the long-term effects of BST on dairy cow performance ................................................................ 24

## 4. Influence of BST on the health, reproductive performance and welfare of lactating dairy cows ........ 27

Evaluation of the general health status of animals under prolonged treatments with BST ........ 27
Influence of BST treatment on reproduction ........ 30
The influence of BST treatment on the welfare of cows ........ 32
Farm needs and requirements for the correct treatment of cows with BST ........ 33
Influence of cow breeds on response to BST treatments ........ 35

## 5. Characteristics of the milk produced by BST-treated cows ........ 37

Influence of BST treatment on the carbohydrate, fat and protein composition of milk ........ 38
Influence of BST treatment on the ion, enzyme, vitamin and hormone composition of milk ........ 42
Chemical and chemicophysical characteristics of the milk produced by BST-treated cows ........ 43
Processing properties of the milk produced by animals treated with BST ........ 44
General remarks on milk characteristics and processing properties ........ 45

## 6. Changes induced by BST treatments in the plasma and milk concentrations of somatrotropin and somatomedins ........ 47

Physiological changes in blood levels of somatotropin and IGF-I during growth and lactation ........ 47
Changes by BST in cow blood levels of somatotropin and IGF-I ........ 49
Changes in the somatotropin and IGF-I concentrations in the milk from BST-treated cows ........ 50
Potential infective or immunogenic risks to humans from milk from BST-treated cows ........ 52

## 7. Characteristics of the meat of BST-treated bovines ........ 55

Effect of BST on the chemicophysical characteristics of the meat ........ 55
Residues of BST present in meat after treatment with prolonged-release BST ........ 56

The fundamental differences between human and bovine somatotropin ................................................................. 58
Possible biological activity of BST in humans ................. 58
The BST "chymotrypsin core" ............................................. 60
Relationship between BST and human digestive tract ...... 61

## 8. Biological and pharmacological manipulations alternative to the use of BST ........................................................ 63

Biological manipulations and the practical use of biotechnological products ..................................................... 63
Alternative manipulations by treatments modifying somatotropin levels ................................................................. 64
Alternative manipulations by somatomedins (IGF-I and IGF-II) ............................................................................... 66
Alternative manipulations by antibodies ........................... 68
Alternative manipulations by β-agonists ........................... 69
Alternative manipulations by somatotropin fusion genes ...... 70

## 9. Final remarks on BST activity ............................................ 71

Galactopoietic effect and nutrient partitioning ................... 71
Influence on cow health, welfare and reproduction ........... 72
Influence on milk characteristics and processing suitability .. 73
Alterations in blood and milk somatotropin concentrations. 73
Alterations in blood and milk IGF-I concentrations .......... 74
Residual concentrations in meat ........................................ 74
Influence on meat characteristics ...................................... 75

## 10. Essential bibliography ...................................................... 77

## 11. Subject index ..................................................................... 105

# FOREWORD

It is hardly surprising that there are in current use in the European Community over forty definitions of biotechnology. At its simplest, biotechnology is a scientific process which employs biological agents to work on a material to produce something different; the making of cheese from milk or beer from barley are often quoted examples. The latest developments in biotechnology are however different; we are in the age of genetic engineering. Today science has made it possible to modify the codes of life themselves.

The *new* biotechnology quite rightly causes concern. Human life and the environment around it are fragile. Technologies (and here biotechnology is certainly not alone) which have the potential to upset the relationship between man and the natural environment must clearly be open to scrutiny. Indeed, the fullest possible provision of information and public discussion are crucial. The worries and concerns of the public are legitimate and should not only be expressed but listened to. But also we must realise that terrific benefits for man and the environment may result from biotechnology. We must not therefore dismiss this technology as totally unwanted. To do so would be counter-productive. Instead, informed choices must be made.

This book is a major contribution to the knowledge we have about biotechnology and its applications. It should be read widely. At a time when the need for Europe's technological relance is so clear, politicians, regulators and the public must be able to balance benefits against costs, advantages against disadvantages. Decisions must not be grounded in prejudice; neither should they be avoided.

*Brussels, july 1990*

**Ken Collins**
*President*
*European Parliament Committee on Environment,*
*Public Health and Consumer Protection*

## PREFACE

The growing advances in biotechnology, together with their economic implications, are already matters of public interest. Increasingly, society must make important decisions involving conflicts between products of biotechnology and industrial, political and social concerns. Correct scientific knowledge of the relationship between the products of biotechnology and the human subjects via intermediate agents (animals, environment, etc.) is useful for all citizens, whatever their calling.

The present booklet is intended primarily for either graduate or undergraduate audience and for those seeking their first, perhaps their only, information about the use of bovine somatotropin (BST) to increase the milk yield in cows. A comprehensive book that describes the full body of information at a level that would satisfy the needs of graduate specialists would surely be found intimidating by most readers in their first encounter with the field.

In setting out to prepare this booklet, I kept in mind that the textbooks have a tendency to acquire complex structure and incorporate in the text many tables, figures, sources of data, etc., with the result that they often lose the very clarity of description and organization. Throughout this booklet I have tried to emphasize the basic framework of the BST action rather than encyclopedic detail.

The cascade *biotechnology* → BST → *cow* → *milk and meat* → *humans* requires a multidisciplinary approach. The time has come when a single animal agriculture or biochemical genetics or human pharmacology textbook cannot be all things to all readers. Throughout this booklet there are many interest-provoking items of related information, many dealing with animal physiology, agriculture and food, still others touching on the target point, namely the human health.

In preparing this booklet, I have reviewed all the available references at the time of writing. However, it should be noted that flow of information on all aspects of this complex subject is continuing and accelerating. Furthermore, there are certainly volumes of data on relevant and critical issues which are at this time proprietary information to the Companies investing in biotechnology. These data have not been available for my review at this time.

Finally, I gratefully acknowledge Gianfranca Corbellini for her skilful and patient assistance in editing the text.

In presenting this very simple booklet, I welcome suggestions and criticisms from readers.

*Pavia, july 1990*

**Gianni Benzi**, *MD, PhD*
*Full Professor of Pharmacology*
*Chief of the Department of Pharmacology*
*University of Pavia - Italy*

*Science does not support
the cow with the iron tail.*

# 1

# General biological characteristics of bovine somatotropin (BST)

The somatotropins are single chain polypeptides of about 190 amino acids, produced by the anterior pituitary gland. The amino acid composition in each species is specific so, while the bovine and ovine somatotropins are very similar, the bovine somatotropin differs by about 35 % from the human one.

Bovine somatotropin (BST) is characterized by the ability to increase both the body weight and the uptake of amino acids into protein chains (protein accretion) and, if administered to lactating dairy cows, to increase milk production (galactopoietic properties). These properties have been widely studied since the 70's, when recombinant DNA technology made it possible to obtain sufficient amounts of individual molecular species to assess their effect and mechanism of action in enhancing milk secretion in dairy ruminants.

## RECOMBINANT AND PITUITARY-DERIVED SOMATOTROPINS IN A COMPLEX INTERRELATED NEUROENDOCRINE HIERARCHY

The hypothalamus is the coordination center of the endocrine system : it receives and integrates messages from the central nervous system, producing a number of hypothalamic regulatory hormones for the pituitary gland. Each hypothalamic hormone modulates the production of a specific hormone by the anterior or posterior portion of the pitui-

### General biological characteristics of bovine somatotropin

tary, with stimulatory or inhibitory action. The specific pituitary hormones (e.g., FSH, LH, somatotropin, TSH, ACTH) are carried by the blood to specific receptors located in the cells of the target tissues (Fig. 1.1).

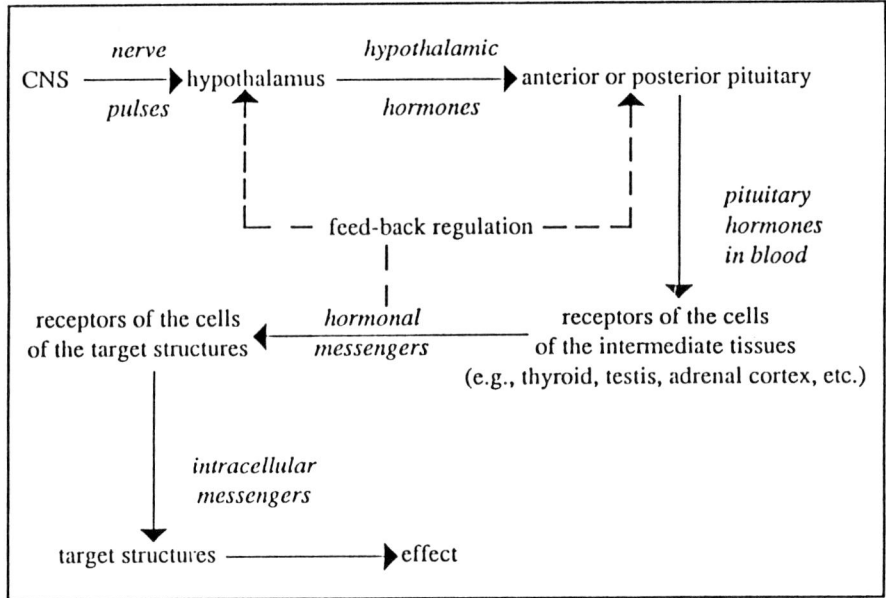

*Fig. 1.1*

Furthermore, the endocrine system is regulated by interconnecting feedback controls. For example, the hypothalamus releases the thyrotropin-releasing hormone (TRH) to the anterior pituitary, causing it to release thyrotropin which, in turn, stimulates the thyroid gland to release the thyroid hormones (thyroxine and triiodothyronine) for the target tissues. The circulating thyroid hormones in the blood inhibit by feedback mechanism the secretion of TRH by the hypothalamus and of thyrotropin by the pituitary. Moreover, the hypothalamic hormone somatostatin inhibits the secretion of TRH, suggesting that the secretion or action of one hormone may be influenced or regulated by other hormones. Thus, a complex regulatory network controls the activities of the different hormones, including somatotropin.

Individual species' somatotropins are not themselves specific entities and from pituitary extracts it is possible to isolate classes of somatotropin-related polypeptides differing in charge, mass and amino- and carboxyl-terminal residue. The biological functions are not necessarily altered by this heterogeneity of composition. Recombinant human or bovine somatotropins are not biologically different from their respective pituitary-derived hormones, and also have equivalent immunological properties and potencies as their respective somatotropins obtained from pituitary extracts. The recombinant bovine product does not seem to be intrinsically more active than pituitary-derived somatotropin.

The molecule tertiary structure plays a pivotal role for the activity of the hormone protein because it characterizes the ability of somatotropin to interact with its specific receptors which are involved in the induction of the biological activity of the hormone. The amino acid sequence of the hormone protein determines the tertiary structure, namely the configuration produced by the folding of the protein chain and the stabilization of the folding by the formation of intramolecular bonds between the amino acids involved.

If the amino acid sequence is not altered in such a way that receptor binding is affected, the activity of the protein is not significantly modified. Recombinant bovine somatotropins are between 0 to 5 % different in amino acid composition from pituitary somatotropin, due to the biosynthetic manufacturing process involving the incorporation of one or more amino acids to one end of the molecule. The absence of biological activity of BST in humans would not be affected because the tertiary structure of the receptor binding domain is not modified. While some recombinant BST products are chemically distinguishable from pituitary BST by amino acid sequencing, they are biologically indistinguishable from pituitary BST.

## INFLUENCE OF BST ON THE NEUROENDOCRINE SYSTEM

The most important modifications produced in the neuroendocrine system by the administration of BST involve both somatotropin itself and insulin-like growth factor I (IGF-I). BST induces an increase in the circulating levels of IGF-I, both in late-lactating cows and in non-lactating animals. The effect of BST on IGF-I level is related to the nutritional state of the animals, as is demonstrated by the fact that plasma levels of IGF-I are reduced : (a) in the presence of high levels of somatotropin in fasting humans, and (b) in cows in a state of energy deficit at peak-lactation.

Prolonged administration of BST does not seem to cause significant alterations in anterior pituitary function, because no significant alterations occur in the plasma concentration of ACTH, prolactin, TSH, LH or FSH. Small increases in the plasma concentrations of thyroxine may be observed, but, in contrast, the levels of glucagon and triiodothyronine are not altered. Small increases in circulating levels of insulin are sometimes, but not always, present during prolonged treatment with BST.

These above-mentioned minor endocrine changes, sometimes observed after BST treatment, are due to the different biological conditions prevailing during the evaluations, e.g., the energy state of the cows, the physiological fluctuations in circulating hormone concentrations, and so on.

BST is an antagonist to the action of insulin in the peripheral tissues. A compensatory mechanism is present in the rat, where the somatotropin stimulates the pancreatic islets to increase the synthesis and release of insulin. However, this double kind of mechanism is not present in the ruminants. On the other hand, an indirect influence of BST treatment upon insulin secretions is modulated by the modifications in the circulating nutrient concentration.

In the liver, adipocytes, fibroblasts, lymphocytes and chondrocytes of various non-ruminants, specific receptors for somatotropin are present. Little information, however, is available on the nature and distribution of these somatogenic receptors in the tissues of ruminants. In bovine liver at least two classes of receptors are present which, because of their high or low affinity to somatotropin, show different functions not yet completely clarified. The production of IGF-I is linked to the presence and function of the high affinity somatogenic receptors.

The type, number and function of somatogenic receptors in the liver is modulated by various factors, including blood estrogen levels and the nutritional state. This last relationship is important from the practical standpoint, because the overall response of the animals to treatment with BST is linked to the relative abundance of functional receptors in the target tissues. Because the nutritional state of the animals makes the somatogenic receptors available, the galactopoietic response of bovines to BST administration is obviously influenced by the kind of breeding and the management of the animals. These factors not only play a role in determining the feeding requirements of the animals, but also influence the biological response to BST treatment.

A very important problem is the specificity of the binding of the receptors in the bovine mammary gland with both bovine prolactin and somatotropin. The bovine mammary receptors bind very well to bovine prolactin, but they show very low binding specificity for bovine somatotropin. This suggests that, in the lactating bovine mammary gland, lactogenic but not somatogenic receptors occur. The absence of specific sites for somatotropin in the lactating mammary tissues suggests that prolonged treatments with BST modify the milk production without acting directly on the mammary epithelium. Thus, the increased milk yield may be mediated by other specific factors, as elucidated in section 1.5.

## INDIRECT INFLUENCE OF BST IN CAUSING THE INCREASED MILK YIELD

BST treatment of lactating bovines immediately induces an increased milk yield, whereas for about 4-6 weeks the voluntary feed intake by the cows does not vary markedly. Therefore in the initial stage of BST treatment, the stored energy sources of the body are utilized. Consequently, during this initial stage, the lactating animals are in a negative energy balance, especially if there is a high extra-milk production. Of course, as the voluntary feed intake increases, the animals get back to a positive energy balance. This indicates that the change in bovine metabolism which makes the extra-milk yield possible must take place during BST treatment.

The effect of BST administration on the concentration of major substrates related to the energy metabolism differs in the various tissues ; for example, the administration of BST increases the uptake of glucose and esterified fatty acids by the mammary tissues, but reduces their uptake by muscular tissues.

This autoregulating mechanism avoids wide variations in the percentage composition of the major constituents of the milk produced by BST administration. Usually the constituents replicate in percentage the physiological values for the milk of untreated animals, apart from a minor change in milk protein and rise in milk fat concentration which may occur.

The altered protein concentration in the milk of animals treated with BST may be related to the modification in the intracellular availability of amino acids for the milk protein synthesis, rather than to a decrease in r-RNA involved in the synthesis of the milk protein, or to a decreased capacity to synthesize and secrete proteins.

The percentage repartition of the different proteins of the milk usually remains unaltered during treatment with BST, with the exception of a possible increase in α-lactalbumin in the milk of BST-treated animals, in comparison with untreated ones. The modifications in both muscle protein accretion rate and amino acid utilization for gluconeogenesis probably permit the sparing action of amino acids for extra-synthesis of milk proteins.

The increase in fat content in the milk of BST-treated animals is due to the strong effect of somatotropin on lipid metabolism. BST decreases the lipid synthesis :

(a) with a sparing effect on acetate and glucose, and

(b) with a consistent increase in the lipolysis and the release of non-esterified fatty acids (NEFA).

The result of this rearrangement of the lipid metabolism by BST is that treated animals have less body fat than untreated ones.

The rise in the milk fat content which may occur in treated animals is related to an increase in the plasma content of fatty acids mobilized from peripheral tissues. This increase in the concentration of the plasma fatty acids is present both in animals with a negative energy balance and in those with a positive one. The fact that the plasma fatty acids increase independently of the positive or negative energy balance, indicates that the BST action on the lipid metabolism of the animals plays a pivotal role.

BST increases lactose output by an increased diversion of glucose to the bovine mammary gland. The lactose level in milk is supported by two key enzymes ($\alpha$-lactalbumin and galactosyl-transferase) and by the availability of metabolic substrates, particularly of D-glucose. During BST administration there is:

(a) an increased lactose synthesis, due to increased utilization of the glucose in the mammary gland;

(b) a decreased glucose oxidation in the peripheral tissues;

(c) an increased hepatic gluconeogenesis from propionate that is converted into glucose by a pathway that occurs in both ruminants and non-ruminants, being quantitatively much more important in the former.

Gluconeogenesis plays a pivotal role on cattle metabolism. Gluconeogenesis from pyruvate, lactate, citric acid cycle intermediates, and from many of the amino acids, proceeds by enzymatic pathways: (a) which differ from the corresponding catabolic pathways; (b) which are independently regulated and require input of free energy. The biosynthetic pathway from pyruvate to glucose takes place in the liver and in the kidney, utilizing eight of the glycolytic enzymes present in large excess and functioning reversibly.

Three irreversible steps in the downhill glycolytic pathway cannot be used in gluconeogenesis and are bypassed by alternative reactions catalyzed by quite different enzymes:

(1) the conversion of pyruvate into phosphoenolpyruvate, via the formation of oxaloacetate;

(2) the dephosphorylation of fructose 1,6-diphosphate;

(3) the dephosphorylation of glucose 6-phosphate.

For each molecule of D-glucose made from pyruvate the terminal phosphate groups of four molecules of ATP and two of GTP must be used.

In the cow the rumen represents a large fermentation chamber in which different bacterial species degrade major plant components (particularly cellulose) which are not hydrolyzed by any of the normal digestive enzymes secreted by animals. Thus, the rumen bacteria both hydrolyze the cellulose to yield free D-glucose and ferment nearly all the glucose to form lactate, acetate, propionate, butyrate, etc..

Cows need blood glucose both to supply the tissues with fuel and to furnish the precursor of milk lactose if they are lactating. Circulating glucose concentration tends to increase during BST treatment (even if it is not constant) as it represents the balance between glucose availability and its utilization at the peripheral level.

Cattle are constantly dependent upon gluconeogenesis, which proceeds at a very high rate in the bovine liver. Lactate formed in the rumen by bacterial fermentation is absorbed into the blood and converted into glucose by the liver.

Another major product of glucose fermentation in the rumen is the propionate which is converted into glucose by a pathway that occurs in both ruminants and non-ruminants, being quantitatively much more important in the former. In fact, the conversion of propionate to glucose is much slower in nonruminants, where propionate is produced only during the oxidation of fatty acids and during the oxidative degradation of some amino acids (i.e., methionine and valine).

Finally, it should be stressed that cows secrete much urea from the blood into the rumen, where microorganisms use the urea as a source of ammonia to manufacture amino acids, which are then adsorbed and utilized by the cows. The overall simplified equation of the urea cycle (involving ornithine, citrulline, arginino-succinate and arginine in reactions catalyzed by enzymes distributed between the mitochondria and the cytosol) is :

$$\text{urea} + 2ADP^{3-} + 2Pi^- + AMP + PPi^{3-} + H^+ \rightarrow$$
$$\rightarrow 2NH_4^+ + HCO_3^- + 3ATP^{4-} + H_2O \quad (1.1)$$

Thus one urea molecule produces two amino groups and $HCO_3^-$. Without the aid of rumen microorganisms, cows cannot by themselves exploit urea as a source of amino groups for amino acid synthesis, since they lack the specific enzymes.

## BIOLOGICAL MODIFICATIONS BY BST TREATMENT

The metabolic changes described in sections 1.2 and 1.3 are related to the lactation period during BST treatment. During peak-lactation, BST administration to cows in negative energy balance induces an increased plasma glucose concentration, without any particular modification in the acetate and urea concentrations, even though there is a small increase in their irreversible losses. In this condition of peak-lactation, BST causes no significant alteration in the plasma concentration of non-esterified fatty acids (NEFA), although there is a decrease in their irreversible loss.

In cows in mid-lactation, treatment with BST increases the plasma concentrations of both glucose and NEFA, this being accompanied by a decrease in the plasma concentration of acetate and urea. There are no irreversible losses of glucose and acetate, while that of NEFA increases.

These differences induced in the metabolism by BST treatment in the different phases of milk yield reflect an increase in the utilization of the fatty acids as metabolic substrates in mid-lactation, together with a sparing action on both glucose and acetate utilization. Instead, at peak-lactation, glucose and acetate appear to be the most important sources of energy for mammary activity to milk yield.

All these biological observations clearly indicate that the effects of BST administration on the metabolism of the animals and on lactation cannot be discussed taking into account just a single parameter, but must be evaluated in their full complexity. The biochemical variations caused by BST are the result of its modulatory effect on the various metabolic pathways and, in particular, on the biochemical pathways by which the physiological milk yield is achieved.

Since the milk yield requires per se many metabolic rearrangements involving the whole body of the animal, it is to be expected that specific metabolic parameters show specific modifications. These modifications bring about the preferential higher availability of substrates to enhance in the mammary gland the content of nutrients required by the increased milk yield.

Therefore, the BST treatment, through a metabolic action modifying the availability of substrates for the mammary glands (particularly of glucose, acetate and fatty acids), allows an increase in the milk yield. This selective provisioning is consistent with the increase in the mammary arterio-venous differences for glucose and acetate.

The influence of the administration of BST in lactating animals is demonstrated also by the "mammary blood flow/milk yield" ratio. The value of this ratio is about 400 : 1 in cows at peak-lactation. In late-lactation the value of this ratio increases to 700 : 1. If these animals are treated with BST, the ratio returns to the peak-lactation value.

## MODE OF ACTION OF BST IN MILK PRODUCTION

It was previously stated that there are no functional receptors for somatotropin in lactating mammary tissues. Considering that BST administration markedly affects the milk yield, it is thus possible to conclude that BST affects the mammary gland indirectly. This observation is supported by two other experimental observations :
(1) that somatotropin has no effect at all on the mammary tissue when the tests are conducted in vitro ;
(2) that milk secretion is not stimulated when BST is infused directly in the arterial blood flow supplying the mammary glands.

The most important factors to be considered as biological intermediates for the stimulation of the mammary gland during BST treatment are the somatomedins and, in particular, the insulin-like growth factor I (IGF-I). The plasma concentrations of IGF-I are increased three- or four-fold during short-term treatment with BST in lactating dairy cows. Keeping in mind that :
(a) BST increases the mammary blood flow, and
(b) BST enhances 3- to 4-fold the IGF-I plasma concentrations, the total result of the BST action on the mammary gland is a 5-fold increase in the amount of IGF-I perfusing the gland.

Therefore, the increase in the concentration of IGF-I induced by BST not only depends on the circulating blood, but also involves the mammary tissue itself. In fact, a BST treatment, able to increase the milk production by about 10-20 %, increases 3-fold the IGF-I concentration in the mammary tissue taken 24-30 hours after the last injection with somatotropin. At this time the circulating somatotropin is back to its basal value and the IGF-I is still twice as high. So, the levels of IGF-I must be considered expressions of cumulative responses of different mammary and non-mammary tissues.

The galactopoietic effect of BST is consistent with an increase in IGF-I secretion in the milk, even if the peak levels are within the range of the physiological milk levels of IGF-I in the lactating cow. In dairy cows the BST-induced variation in the concentration of IGF-I in the milk slightly precedes the modification in the rate of the milk yield.

At this point the question can be raised whether IGF-I stimulates the activity of pre-existing milk producing cells, or whether it induces an increase in the number of those cells. An answer to this question could be seen from research in lactating goats in which IGF-I is infused arterially into one mammary gland, and the variations in the milk secretion rate are measured in both the infused and non-infused gland.

In the non-infused mammary gland there is an increased milk production of 15 % (due to IGF-I present in recirculating blood), while in the infused gland this increase reaches 30 % (due to IGF-I directly arterially infused). Similarly, the milk produced in the non-infused gland shows an increase of about 15 % in IGF-I concentration, versus an increase of about 80 % in the concentration of IGF-I in the milk produced by the directly infused gland.

This kind of observation leads us to two conclusions :

(1) the action of IGF-I on the epithelium of the mammary gland is on the pre-existing cells, because of the rapidity of its development reaching significant values few hours after the beginning of the intra-arterial infusion ;

(2) the response is much higher in the directly arterially infused gland, suggesting that IGF-I acts directly on the mammary epithelium to increase milk secretion.

This unilateral infusion experiment is performed utilizing the IGF-I in the free form. However, when IGF-I is in contact with the blood it binds to a plasma protein with a molecular weight of 150 kD. The binding of IGF-I with this 150 kD protein :

restricts its movement out of the vascular compartment,

decreases the ability of the factor to transfer from the blood to the peripheral tissues,

diminishes its biological activity.

The unilateral infusion experiment shows that, if after 4 hours from the beginning of infusion the level of IGF-I increases by about 65 %, then only 14 % of IGF-I in the blood is present in the free form.

Therefore, the delayed and lower response to IGF-I in the non-infused gland is caused by the low transfer and efficacy of the recirculating protein-bound IGF-I, which limits its availability for the target gland tissues. In contrast, in the gland tissues directly infused with IGF-I, the factor goes directly to the gland epithelium in its free, non-protein bound form.

IGF-I is anyway able to induce an increase in the blood flow in the mammary gland. Thus, IGF-I not only increases the milk yield by stimulating the gland epithelium, but it also simultaneously increases the mammary blood flow which brings the biochemical substrates necessary for the production of milk nutrients.

These observations suggest that IGF-I may be an intermediate factor responsible for the activity of BST because in the mammary gland it is able to influence both the tissue response to the stimulation, and to change the mammary blood flow : (a) through a direct stimulation of the vascular system, or (b) through an indirect stimulation of that system, by the release of vasodilator compounds produced by the increased mammary metabolic activity.

IGF-I and IGF-II are both lipolytic at low concentrations, and IGF-II seems to be more potent than IGF-I. Thus somatotropin may enhance lipolysis by releasing IGF-I, or by altering the responsiveness of adipocytes to other lipolytic or lipogenic factors, namely insulin and catecholamines. On the other hand, the BST action may result from direct action on adipocytes in the presence, however, of the complete endocrine environment.

## GENERAL REMARKS ON THE BIOLOGICAL ACTION OF BST

BST is able to modify the lactation in dairy cows by increasing the milk yield through complex metabolic mechanisms. The modification induced can be characterized by three effects.

The first effect is the modified utilization of nutrients and their mobilization from non-mammary tissues, with a sparing action for the essential substrates available to milk production. This effect of BST is related to a direct influence of somatotropin on the function of various tissues (particularly liver and adipose tissue), and/or to a modification of the responsiveness of the tissues to other hormones influencing the substrate metabolism, namely catecholamines and insulin.

The second effect of BST is the local increase in the mammary blood flow, although it is not completely clear if this event is the cause or the consequence of the increased milk production. The third effect of BST is the modification of the mammary gland function to increase milk production by the influence of intermediate factors such as somatomedins (IGF-I and IGF-II). The enhanced milk yield by BST occurs when these three events act in unison.

Methionyl-BST (met-BST) is a recombinant BST with a methionine addition to the N-terminus. The molecule was described as more potent than pituitary BST in increasing milk production compared to a pituitary BST. However, the pituitary BST contained a substantial amount of degraded BST, thereby making the standard less active. The effectiveness of met-BST in lactating dairy cows is, in fact, indistinguishable from BST with the same amino acid sequence as a variant produced naturally by the bovine pituitary gland.

# 2

# Influence of BST on the metabolism of the whole animal and of specific organs

As explained in Chapter 1 regarding the general biological effects of BST, the hormone itself is capable of modifying various metabolic pathways, as follows :
— *on carbohydrate metabolism :*
   increase in glucose production ;
   decrease in glucose oxidation ;
— *on lipidic metabolism :*
   decrease in the fatty acids synthesis ;
   increase in fatty acid oxidation ;
— *on protein metabolism :*
   increase in nitrogen retention ;
   increase in protein synthesis ;
— *on mineral metabolism :*
   increase in calcium ion absorption ;
   increase in calcium ion accretion ;
— *on nucleic acid metabolism :*
   increase in cell proliferation.

## THE ACTIVITY OF BST ON CARBOHYDRATE AND LIPID METABOLISM IN THE WHOLE ANIMAL

From a general point of view, somatotropin shows two basic characteristics :

(1) the first characteristic regards the metabolic processes that are partially, but not exclusively, mediated by IGF-I ; and

(2) the second characteristic regards the modification of the nutrient supply, which is most probably related to the direct metabolic action of BST itself.

The amount and the quality of the effects induced by BST are closely related to the physiological condition of the animal. For example, in growing animals prolonged treatment with BST causes an increase in muscle protein accretion, while during lactation BST can induce the opposite effect. Furthermore, the result of treatment with BST depends on the nutritional state of the animal. For example, in animals that remain in a positive energy balance, during treatment with BST it is possible to observe a decrease in lipogenesis of adipose tissue, with no increase in lipolysis. In contrast, if the animals are in a negative energy balance a BST-dependent increased lipolysis occurs.

As already indicated in Chapter 1, if the administration of BST is not immediately balanced by an increase in food intake, the animals go into a transient phase of negative energy balance. This event results from the BST effect that causes a re-arrangement of the metabolism elsewhere (eumetabolic or homeorhetic action) in the lactating cow body, to provide the nutrients that the mammary gland needs for increased milk production. This event occurs naturally during early lactation in untreated animals.

This kind of metabolic re-arrangement is also supported by the fact that BST does not induce any direct effect on the appetite or on the carbohydrate, protein and lipid digestion. This is true in both growing and lactating animals. Therefore, the metabolic effect of BST automatically leads to an increase in the food intake as a secondary response, without modification of feed desire at the brain level, or of digestive capability at the gastric level.

A further indirect demonstration of the eumetabolic action of BST is represented by the increased absorption of calcium to make the ion available to the mammary gland, because of the increased milk yield. As a matter of fact, it is dangerous to decrease the calcium availability for net bone growth. So, the most correct biological response is to increase the calcium absorption from the diet.

The increase in milk production, as a response to BST treatment, needs an additional increase in the amount of glucose available in the mammary gland for lactose synthesis. In this condition the irreversible loss of glucose can either increase or remain unmodified.

These different responses are not surprising because glucose must be evaluated not in absolute terms, but rather in terms of balance between availability and use (glucose turnover). The higher utilization of glucose for lactose synthesis in the mammary gland can be balanced by a decrease in muscle glucose oxidation to carbon dioxide, concomitant with an increase in the oxidation of non-esterified fatty acids.

During treatment with BST there is an increase in the irreversible loss of non-esterified fatty acids, both in lactating cows and in growing heifers. The increase in fatty acid availability as energy substrate causes a decrease in glucose oxidation in the peripheral tissues, perhaps through a modification of pyruvate dehydrogenase activity. All these modifications are linked to the metabolic re-arrangement that develops in the initial period of BST treatment, when the needs of mammary production are increased by the treatment itself.

After this transient initial phase when milk production is increased without any increase in food intake, the metabolic adaptation takes place through the intervention of nutritional factors. This is demonstrated, for example, by the fact that the plasma concentration of non-esterified fatty acids remains at normal level in animals treated for a long period with BST. These facts demonstrate again the lack of extensive variations in the oxidation of fatty acids in subsequent phases after the initial period of unbalanced energy state.

## BST INTERACTION WITH THE ENDOCRINE SYSTEM

In lactating animals the administration of BST increases the plasma level of IGF-I, usually without any increase in the plasma concentration of insulin. Furthermore, treatment with BST induces no apparent effect on the plasma concentration of a number of hormones, such as prolactin, glucagon, hydrocortisone, thyroxine, triiodothyronine, etc.. Treatment with BST decreases the insulin ability to enhance the utilization of glucose in the peripheral tissues. This inhibitory effect of BST on peripheral insulin activity facilitates the preferential utilization of glucose by the mammary glands.

## EUMETABOLIC (HOMEORHETIC) EFFECT OF BST

The increase in glucose irreversible loss results in increased glucose production, especially by the liver. In the transient initial period of BST treatment, this increase is due to an augmented glycogenolysis, and subsequently to an increased gluconeogenesis. The liver has specific receptors for BST, and the BST effect on glucose metabolism is caused directly by BST itself and does not require any IGF-I mediation.

While carbohydrate metabolism in the liver may be largely defined and predicted, the modifications by BST of hepatic lipid metabolism are not so clear and foreseeable. Treatment with BST increases the plasma level of free fatty acids, with an increase in the uptake of these fatty acids which may be :

(a) esterified and secreted as lipoproteins, or

(b) transferred into the mitochondria to be oxidized or converted to ketones, and released into the blood for tissue metabolic utilization.

BST treatment increases the fatty acid oxidation at the mitochondrial level, without a consistent increase in the formation of ketones (acetoacetate and $\beta$-hydroxybutyrate). In fact, the administration of BST does not induce any variation in the cow's plasma concentration of ketones, suggesting a lack of effect on ketogenesis. In any case, the modification of the lipid metabolism in the liver is evident in the transient initial period of negative energy balance, when the metabolic effect of BST is operating and resulting in increased milk production without any consistent increase in food intake.

Long-term treatment with BST decreases the amount of peripheral fat tissues, as a result of decreased synthesis, and/or increased lipolysis, or of both mechanisms, so the animals tend to be lean. The fatty acids required for lipid synthesis may be produced through lipogenesis or released from plasma lipoproteins by the action of the lipoprotein lipase. In the fat tissue, long treatments with BST decrease the amount of lipogenesis.

The mechanism whereby BST increases lipolysis has not been fully elucidated. Probably the lipolytic effect may be mediated by the somatomedins and/or by the altered responsiveness of adipocytes to either catecholamines or insulin. In fact lactation results in both an increase in the number of adrenergic receptors of adipocytes in cattle, and an enhancement of their sensitivity to the catecholamines. On the other hand, BST modulates the stimulation of insulin on either lipogenesis or lipid esterification.

## EFFECTS OF BST IN LACTATING ANIMALS

The somatotropin stimulates the milk yield and, as a consequence, the synthetic processes at mammary gland level are increased with an increase in the uptake of nutrients needed for milk production. Furthermore, the somatotropin increases the blood flow to the mammary gland, and increases the nutrient supply for the synthetic processes. Thus the nutrient uptake must be increased and this is a key part of the BST mechanism of action.

As previously indicated, the BST mechanism for carbohydrate utilization is very clear, but the effect on the metabolism of fatty acids is not so clear and unequivocal. It is known that the free fatty acids and the triacylglycerols of the blood can be taken up and utilized in different metabolizing processes.

This is related to the fact that lipids have different pathways of utilization inside the body, depending on the physiological state. Thus, it is difficult to define lipid metabolism only as it applies to milk production, because during BST treatment the animals are in various physiological and nutritional states. The BST eumetabolic action will vary according to the energetic needs of organs and tissues during the different periods of lactation, and thus the lipids are more or less preferentially involved in the various metabolic pathways.

Treatment with somatotropin increases the proportion of mammary carbon dioxide produced from fatty acids in comparison with that derived from glucose oxidation, which does not significantly change. On the other hand, a large part of the carbon dioxide deriving from the utilization of glucose is related to the pentose phosphate pathway, rather than to the classic tricarboxylic acid cycle. The BST-increased oxidation of fatty acids may induce a switch from lactate uptake to lactate output by the mammary gland. This switch of the oxidative process through a larger fatty acid use causes a decrease in the value of the mammary respiratory quotient.

Thus, BST action on the mammary gland results in :

(a) an increased uptake of precursors (namely, glucose, fatty acids and amino acids) concurrently with the increased synthesis of milk constituents, and

(b) an increased utilization of extra-mammary fatty acids,

both for oxidation and for milk lipid synthesis.

## GENERAL REMARKS ON THE METABOLIC EFFECT INDUCED BY BST

The effects of BST on mammary metabolism are thought to be indirect, because :
in cattle it does not appear that any receptors for somatotropin exist in the mammary epithelial cells,
the addition of BST in culture of mammary tissue does not change the cellular metabolism.

The BST activity may be mediated by the presence of receptors for the IGF-I that is released by the BST itself. Anyway, this explanation relates only to the mechanism by which the increase in milk production under treatments with somatotropin becomes evident.

The period in which there are significant metabolic changes in lactating animals is the first period of adaptation to BST treatment, promoting milk production with no concomitant increased food intake. In this phase it seems that the bio-metabolic situations are similar to those found in genetically high-yielding cows. During this first period of BST treatment the animals tend towards a negative energy balance, with lipid mobilization and utilization. Subsequently, the eumetabolic action of BST tends to compensate this energy imbalance.

During the initial period of BST treatment the use in extra-mammary tissues of fatty acids as metabolic substrates is very important, because it allows the reduction of glucose utilization by muscle and other tissues, leaving the glucose available to the mammary tissue for milk synthesis. During subsequent periods, prolonged BST treatment increases the food intake, balancing the extra-need for nutrients and avoiding the necessity of mobilizing fatty acid reserves.

In this way, the eumetabolic action of somatotropin maintains the energy balance in a steady-state, because the animals can find from extra-dietary nutrients all the substrates necessary both for the maintenance of the metabolic equilibrium and for the higher metabolic need for increased milk production.

The eumetabolic action of somatotropin results both in a decrease in lipogenesis in the adipose tissue and in insulin-resistance in extra-mammary tissues, resulting in decreased glucose oxidation by peripheral tissues. For lactating cows, the advantage of this glucose sparing action is clear.

In summary, we cannot expect a single action of BST because of its more complex and sophisticated modulatory action as a function of the biological and physiological needs of the body as a whole.

# 3

# Long-term effects of administration of BST in dairy cattle

The data reported in the preceding sections generally relate to the action of somatotropin on the energy metabolism of the cow. From a practical point of view, it is important to take into account the long-term effects in dairy cows treated with BST, with particular reference to milk production, milk composition, energy balance of the animal, body weight, health, nutrition, feed efficiency, reproduction, and so on. This must be done in relation to both the BST dosage administered and the lactation period.

## BST LONG-TERM EFFECTS ON MILK PRODUCTION AND COMPOSITION

The extra-amount of milk produced by daily treatment with BST is a dose-dependent response. For example, for treatments of mean duration of 32 weeks (starting from 5 to 13 weeks after calving) with a daily dose of 5 mg of BST it is possible to observe an increase of about 10 % in milk production in comparison with the control cows. The administration of BST at a dosage between 30 and 50 mg per day induces an increase of about 20 %. In both cases the response to BST is very rapid and the maximum effect is already evident in one week or less.

These data are related to daily injection treatments : BST preparations are, however, available which permit the slow release of BST over a long period of time. This ensures better animal welfare and facilitates the practical use of this substance without the need of daily intervention for long periods of time. From Eli Lilly, Monsanto, American Cyanamid, etc. there are available *prolonged-release preparations* whose excipients are still proprietary information. Moreover, the BST dosage and rate of administration differ between Companies.

The prolonged-release preparation from Eli Lilly is used at 320, 640 and 960 mg and injected every 28 days, corresponding theoretically to 11, 23 and 34 mg BST/day, with production responses of 56, 60 and 75 % of those obtained by daily injections of comparable amounts. The BST from Monsanto is used at 500 mg injected every 14 days, corresponding theoretically to 36 mg BST/day, with a production response of 80 % of that obtained by daily injections of comparable amounts. The BST from Cyanamid is used at 350 or 700 mg injected every 14 days, corresponding theoretically to 25 or 50 mg BST/day.

The various prolonged-release preparations increase milk production, while overall composition remains quite stable. The fat content in the milk can increase along with the increase in the milk yield, which is probably due to body lipid mobilization when milk production is at its peak. The long-chain fatty acid or lactose contents of the milk may show cyclic variations between two injections.

## VOLUNTARY DRY MATTER INTAKE AND CALCULATED ENERGY BALANCE IN ANIMALS DURING NORMAL LACTATION AND WHEN BST TREATED

The nutrient supply is unrelated to the cow requirements at any period of the lactation cycle. In fact, in early lactation the requirements increase more rapidly than the feed intake capacity ; thus, the body reserves of lipids, proteins and minerals are mobilized and used to meet the deficits. These reserves, however, must be replenished before the next calving.

During the first 2 months of lactation, the *lipids* mobilized usually vary between 15 and 60 kg, depending on the level of production. If a large energy deficit extends beyond the third month of lactation, actual reproductive difficulties and low persistency of milk yield occur. The deposition of 30 kg lipids (corresponding to 40-45 kg live weight) cannot take less than 2 months in mid-lactation and high producing cows may need as long as 4 to 5 months to replenish their body reserves.

In contrast, the capacity for *protein* mobilization is much more limited, lasts for only a short time (2 to 5 weeks), and is less than 15 kg in underfed cows in early lactation : more than half of these proteins originate from muscle and the remaining part from viscera and organs. Without reducing milk production, in high producing cows, the total protein mobilization cannot exceed 5 to 10 kg, i.e. the equivalent of the amount of protein excreted in 100 to 200 kg milk.

When intake is below the steady-state requirement (where the cow neither gains nor loses body weight), the dairy cow uses energy from its body reserves to produce milk, and when the intake exceeds requirements, part of the extra energy is deposited in body reserves. When cows are seriously under-nourished there is a risk of impaired reproductive performance.

The voluntary dry matter intake is minimal at calving and subsequently increases less rapidly than milk production until it peaks during the third month of lactation. The intake is equivalent to 60-80 % of this peak during the first week of lactation, but rises to about 95 % as early as the second month, being dependent on many factors (e.g., milk production, lactation number, forage quality, animal health).

If concentrate is supplied individually according to milk production, the voluntary dry matter intake of maize silage increases between the first and the 8th week postpartum. The corresponding increase in forage voluntary dry matter intake is much less with grass silage and hay.

After its plateau value, voluntary dry matter intake decreases in parallel with milk production and concentrate supply. Forage voluntary dry matter intake does not increase when the concentrate supply is reduced and, in these conditions, forage does not compensate for lack of concentrate. There is a further marked decrease in forage voluntary dry matter intake during the last month of lactation and especially during the dry period.

Treatment with prolonged-release preparations of BST produces an increase in the intake of nutrients after the first adaptation period. In fact, the amount of extra-feed ingested corresponds quite faithfully to the theoretical amounts that can be calculated for untreated cows increasing their milk production by the same amounts as those induced by BST treatment. This agrees with the view of some authors who state that the treated cows are in some ways similar to genetically higher yielding cows.

The calculated energy balance tends to be negative during the early lactation phase, whereas it tends to be in equilibrium during the subsequent phase, when there is a higher feed intake. This is in agreement with the observations that animal body weight tends to decrease during the first periods of BST treatment and to stabilize during the subsequent periods.

A lower body weight in cows during the first period of BST treatment occurs, this event being more or less reversed during the following period, as a result of both the energy balance trend and the lipid mobilization during peak-lactation and the re-deposition of lipids during lactation when milk production falls.

The changes in fat metabolism depend on the nutritional status of the animals, on the season when the treatment is given and the concentrate usage. All these variations on the energy balance, body weight, body composition and body lipids in a particular lactation phase are supported by the eumetabolic action of somatotropin, as described in Chapter 2.

All the calculated modifications in dry matter intake and energy balance vary in relation :

(a) to the different conditions in which treatments are given ;

(b) to the different seasons when the treatments are given ;

(c) to the caloric contents of the feed, in function of its storage and conservation ;

(d) to different methods used in estimating the energy density of the diets ;

(e) to different kinds of concentrate put at the disposal of the animals.

The traditional concentrates are based on cereal grains and supply a large amount of starch that, with or without BST treatment, could depress rumen pH, cellulolytic activity of the rumen microbial population, forage intake and digestibility, and milk fat percentage. These risks can be reduced by replacing certain by-products rich in non-lignified cell wall (such as beet pulp) which ferment slowly in the rumen whilst being highly digestible.

Wheat bran allows a greater forage intake than cereal grains. Saturated animal-fats or blended animal-vegetable fats are added to obtain beneficial effects (particularly when the roughages have a low lipid content) but, with or without BST treatment, they can depress milk fat and protein content on maize silage diets.

By giving feeds rich in rumen-undegraded dietary protein, the amino-acid requirement of the cow can be provided in early lactation although the supply of microbial amino acids is still limited by the low feed intake capacity of the cow. It is suggested that feeding rumen-protected meals also seems to be a way of supplying the appropriate quantities of the essential amino acids (such as lysine and methionine) which are the most limiting with maize silage diets.

## RELATIONSHIPS BETWEEN RESPONSE TO THE BST TREATMENT AND MANAGEMENT CONDITIONS

Differences in milk production response to BST between heifers and cows may sometimes be present. However, because of the different conditions in which the observations are being taken, it is difficult to provide a clear picture of the parity distribution in a herd. In fact, in some cases productive response is lower in heifers than in cows, but in other cases it is higher.

The increase in milk production by BST treatment shows an individual variation that cannot be related to the milk production potential and that cannot be forecast as a function of the parents of the animal itself. In comparison with untreated cows, the percentage of increase in production response to BST decreases when the milk yield increases.

The lactation stage, at which the treatment begins, does not influence in any particular manner the production response. For a given initial stage of stimulation of lactation, the persistency of the response to BST during the remainder of the lactation is quite variable. It often tends to decrease, particularly in relation to management conditions, pregnancy stage or the physiological needs of the cows to maintain body reserves. On the other hand, the milk production response to treatment with BST during a second consecutive lactation is of the same magnitude as during the first lactation.

Because the influence of the diet on the response to BST has been evaluated in different experimental conditions, the results are not consistent. It seems, however, that higher responses could be obtained with several complete mixed rations of different energy content given according to the milk yield.

It is necessary to increase the supply of nutrients to animals under BST treatment to meet the increased metabolic requirements related to higher milk secretion. Cows receiving low concentrate diet weigh less at the end of BST treatment, and they have a lower persistency during the next BST lactation.

Increasing the percentage of rumen-undegradable protein does not increase the response to BST in cows of lower milk potential. A normal diet qualitatively balanced is more than enough to provide the increased metabolic requirements due to increased milk production. Cows make use of microbial action to predigest grasses and leafy plants. To avoid vitamin $B_{12}$ deficiency in bacterial flora, cows have a high nutritional requirement for cobalt. In regions where the soil is poor in cobalt, cobalt deficiency in cattle is a serious problem.

If the environmental situation is maintained in terms of good management and balanced nutrition there are no particular alterations in the response to treatment with BST. Furthermore, the animals do not suffer any gross harm as a result of receiving an intramuscular or subcutaneous injection of BST every 14 or 28 days.

In the practice of treatment with BST, the harm which could occur to people taking care of the animals is practically nil. The possibility that the veterinarian injects the BST into one of the personnel or that members of personnel inject themselves is very remote and inadmissible in normal working conditions. However, even were the BST to be mistakenly injected into humans it should be stressed that BST has no effect in humans because of its 35 % difference from human somatotropin (see also sections 7.3 and 7.4).

The possible problem of the influence of injections of BST on the welfare of the animals, of the veterinarians and of those who take care of the animals will also be discussed in Chapter 4.

## GENERAL REMARKS ON THE LONG-TERM EFFECTS OF BST ON DAIRY COW PERFORMANCE

The treatment with prolonged-release preparations of BST has a high potential for increasing milk production, with an increment of about 1000 kg for a period of 200 days of lactation. These responses, however, can be lower, especially in European feeding conditions. Therefore, the increase in milk production may be from less than 5 % up to 25 %, depending on both the initial yield level of cows and the duration of the treatment itself.

During the first period of treatment with BST, a negative energy balance and a decrease in body fat reserve in the animal can be observed. This condition increases the delay in conception and in the inter-calving interval, as can be observed in untreated higher milk-yield dairy cows. The initial imbalance of energy becomes balanced in the following periods, both by eumetabolic action of somatotropin and by increased feed intake.

The increase in feed efficiency (calculated as the ratio : fat-corrected milk/dry matter intake) depends :

(a) on the dilution of maintenance requirement in the total requirement, because of increased milk production ; and

(b) on a lower use of nutrients for body tissue deposition relative to milk secretion.

Whether the treated subjects are heifers or multiparous cows seems to have no particular importance in the variation in milk production after the administration of BST.

Regarding the quality of the ration to be supplied to the animals, if this is balanced and meets the needs of the animals there are not any particular problems. There could, however, be a decrease in the body weight of animals fed on forage of poor quality caused by the environment or weather. Therefore, feeding management plays a pivotal role in milk secretion, as well as in improving body condition and reproductive performances of the cows treated with BST ; exactly as it does in cows in high producing herds without BST.

# 4

# Influence of BST on the health, reproductive performance and welfare of lactating dairy cows

As already mentioned in Chapter 3, of particular importance are the health, reproductive performance and welfare of dairy cows under prolonged treatment with BST, even if this is achieved by one injection every 14 or 28 days.

## EVALUATION OF THE GENERAL HEALTH STATUS OF ANIMALS UNDER PROLONGED TREATMENTS WITH BST

*Clinical examination* shows that the body temperature and the respiratory rate are not modified by prolonged administration of BST, although there is a slight increase in heart rate with high doses.

In housed cattle it is common to find *fluid filled bursal swelling* of the carpus and tarsus. This picture is frequent enough and normally it resolves itself when the animals go to pasture. Treatment with BST could increase this synovial symptomatology, especially in animals that are housed for a long time. The increase in the incidence of lameness related to BST manifests itself not in the first phase of lactation, but in the second, in animals housed for more than one year. BST treatment does not necessarily cause an increased incidence of lameness, but probably makes more evident an alteration already common in the cattle, which is probably more closely related to individual farms and management practices.

It is important to establish whether the long-term treatment with BST induces any alterations in the *growth of the bone* and in the *general development* of the animals. The development of the bone in BST-treated animals remains within the normal physiological limits and there is no bone depletion of calcium or phosphorus.

The fact that BST induces an increase in milk production leads (as described in Chapters 1 and 2) to a reorganization of the general metabolism of the animal. According to some authors, animals treated for a long time with somatotropin present a similarity to genetically superior dairy cows for milk production. In these cows the feed intake seems to be genetically adjusted to yield more milk, but their characteristics are markedly influenced by the level of management.

A specific problem is that the metabolic reorganization may cause the appearance of *ketosis, milk fever* and *burn-out*.

In cows, the glucose is used for milk production principally in the synthesis of lactose and of glycerol phosphate, and is utilized in the formation of milk triacylglycerols. When food quality is poor, milk production decreases and the animals sometimes develop ketosis, a condition in which the concentration of the ketone bodies (acetoacetate, D-β-hydroxy-butyrate, and acetone, from partial oxidation of fatty acids) is abnormally high in blood, tissues, and urine.

It is the diversion of glucose and its precursor oxaloacetate to milk production under conditions of extensive fatty acid catabolism which results in ketosis. In this case, propionate plays a pivotal role in averting ketosis. In fact propionate can be converted to succinyl-CoA, through the following reactions :

propionate + ATP + CoA → propionyl-CoA + AMP + PPi        (3.1)

propionyl-CoA + ATP + $CO_2$ + $H_2O$ →
$\qquad$ → D-methyl-malonyl-CoA + ADP + Pi        (3.2)
D-methyl-malonyl-CoA → L-methyl-malonyl-CoA        (3.3)
L-methyl-malonyl-CoA → succinyl-CoA        (3.4)

The last reaction (3.4) is catalyzed by an enzyme containing a tightly bound coenzyme form of vitamin $B_{12}$ : deoxy-adenosyl-cobalamine, in which the cyano group CN is replaced by the deoxy-adenosyl group.

Cows can readily transform succinyl-CoA into oxaloacetate, thus averting ketosis:

$$\text{succinyl-CoA} + \text{GDP} + \text{Pi} \rightarrow \text{succinate} + \text{GTP} + \text{CoA} \quad (3.5)$$
$$\text{succinate} + \text{FAD} \rightarrow \text{fumarate} + \text{FADH}_2 \quad (3.6)$$
$$\text{fumarate} + \text{H}_2\text{O} \rightarrow \text{L-malate} \quad (3.7)$$
$$\text{L-malate} + \text{NAD}^+ \rightarrow \text{oxaloacetate} + \text{NADH} + \text{H}^+ \quad (3.8)$$

The keto-acidosis could be due to modifications induced by somatotropin both in the insulin activity and in the fatty acid metabolism. With the exception of one finding, prolonged administration of BST does not seem to produce ketosis, or milk fever, or a subclinical condition of sensitiveness to the ketosis itself. In BST-treated animals the blood β-hydroxybutyrate level remains in the normal range of values.

The response to BST depends strictly on the *nutrition and management conditions*: if these conditions are optimal, the responses to BST are optimal also. If the nutrition and management conditions are below optimum, the responses to BST are also low, and if the conditions are bad the BST has practically no effect.

The response to BST depends on the *phase of lactation* examined. For example, if between the 9th and 17th weeks of lactation the increased milk production is around 4.5 kg/day, later this increment is reduced to about 3.5 kg/day between the 26th and 33rd weeks, and to around 2.5 kg/day between the 34th and 41st weeks. Thus, the BST response follows the natural production decrease towards the physiological end of lactation.

An important indication of the general health status of the animal is furnished by the *blood chemistry*. With doses of BST similar to those for practical application (around 20 mg/day) the blood profile is not altered, while with higher dosages (for example around 40 mg/day) there is a decrease in the hemoglobin concentration and in the hematocrit value.

Similar comparisons of blood parameters can be made after the administration of prolonged-release preparations: 500 mg of BST every 14 days. In one case, there was a decrease in the hemoglobin content and in the total number of circulating red cells. Thus, in the animals treated with BST, the number of red blood cells decreased from $5.9 \times 10^{12}/L$ to $5.5 \times 10^{12}/L$, a 7 % decrease. It should be noted that these values fall well within the normal physiological limits for dairy cow red blood cells of 5 to $9 \times 10^{12}/L$.

A tendency to blood cell decrease could anyway reflect an alteration in the physiopathological balance between the total number of red cells and the total volume of plasma. Possibly the hematological alteration could be a form of false anemia from a plasma dilution due to a changed metabolic state in animals treated to obtain higher milk production. In any case the precise nature of this hematological alteration requires clarification.

The *somatic cell count* in milk is generally considered as a marker of both udder health (not shown by overt clinical symptoms) and the presence of subclinical mastitis. Long-term treatment with BST either causes no variation in the somatic cell count (9 studies out of 12), or induces a non-significant increase within the normal limits (1 study out of 12), or causes a significant increase (1 study out of 12), or induces an increase which is significant only on two of eight sites (1 study out of 12).

It should be stressed that the somatic cell count is part of the milk quality control system in the whole European Community and, therefore, this parameter should be checked carefully under European conditions of cow management.

*Mastitis* represents one of the most widespread "professional" diseases of dairy cattle and has a high negative economic influence on the costs of the dairy industry. The incidence of mastitis in animals treated with BST seems to be the same as that in untreated animals. Only a few epidemiologic studies indicate a high incidence of mastitis in groups of animals treated with BST ; nevertheless, in these studies the overall incidence of clinical mastitis seems to be notably higher before the treatment with BST.

Regarding both the somatic cell numbers and the incidence of mastitis, the variation in the responses to BST is related also to exogenous factors, such as management, diet, lactation period, milking machine maintenance and the environmental conditions of the farm.

## INFLUENCE OF BST TREATMENT ON REPRODUCTION

Reproductive performance is very important in animals producing meat or milk, because a decline in reproductive efficiency can cause large economic losses.

As already stated, BST administration produces a rapid increase in milk production, while at least 4-6 weeks are needed before the increase in the food occurs. This causes a negative energy balance during the first period of treatment with BST.

Generally, all the exogenous factors that diminish the reserves of the animals cause a decrease in the reproductive performances. Therefore, the management, the quality of the nutrition and the types of nutrients have great importance, not only with regard to the reproductive response to BST treatment, but more notably with regard to the reproductive performance of the animals independently of any pharmacological treatment.

Daily dosages of 10 or 20 mg of BST do not cause any particular alterations in the percentage of days open and pregnancy rate, but more elevated dosages (40-50 mg/day) influence these parameters adversely. There are, however, conflicting data again regarding the days open in animals treated with low dosages of somatotropin. These differences could be related to the environment and nutritional status rather than to the biological characteristics of BST. In any case, with elevated BST dosages utilized for long periods of time, days open increase and the pregnancy rate decreases.

BST treatment with 500 mg every two weeks (beginning two or three months after calving) appears to have no significant effect on reproductive performance. In the reproductive performance of BST-treated cows, the percentage concentrate in the ration of animals plays an important role in the first stage of lactation (1-12 weeks), in the mid-lactation (13-26 weeks) or in the late lactation (27-44 weeks).

In the various periods of lactation, the decrease in the percentage concentrate in the ration causes a decrease in the reproductive performance of the animals, because the energy balance tends to be negative. If a decrease in percentage concentrate in the ration is combined with BST treatment (at the dosages normally utilized to increase milk production) the percentage pregnancy rate is reduced, in comparison with the untreated animals. These data are not surprising, because they confirm that the events causing a negative energy balance become determining factors in reproductive performance.

A generally accepted index of the reproductive performance is represented by the above-mentioned *days open*. An increase in one day open is associated with a 100 kg increase in 44-week lactation yield. It is generally accepted that high-yielding cows are more difficult to get in calf. Indeed, the higher milk producing animals present a higher number of days open : so there is a direct correlation between the increase in milk production and the increase in days open.

In animals treated with BST both phenomena proceed in parallel ; therefore, an increase in milk production corresponds to an increase in days open : exactly as happens in untreated high-yield animals. Thus, the increase in days open due to BST (and so the reduction of the reproductive performance) seems to be related more to the metabolic modification due to increased milk production than to a direct action of BST on the reproductive performance.

In some conditions, BST treatment causes an increase in *twin calves*, with an increase in the incidence of twinning of about 3 % in comparison with untreated animals. Because this is not economically advantageous to the farmer, if this direct link between BST administration and twin calving is confirmed, it would be preferable to start BST treatments only after most of the cows had conceived.

The *fetus* presents no particular harm during treatment with BST, because calves born to cows treated with BST show a body temperature, heart rate, respiratory rate, body weight and other general characteristics no different from those of calves born to untreated cows. Also, the body development of calves born to pregnant cows treated with BST seems to be within the normal physiological limits, and the calves mature normally and show normal reproductive performance and milk production ability.

## THE INFLUENCE OF BST TREATMENT ON THE WELFARE OF COWS

Treatments over very short periods of time do not cause any modification of the welfare of the animals. Several questions, however, can arise over the side-effects of treatment for long periods.

Firstly, these could be due to the lack of suitable equipment and technical ability for performing the BST injections on the various farms. Also, BST administration requires direct veterinary control. On modern farms there is usually continuous adequate veterinary supervision for the normal routine control of the reproduction, health and nutrition of the animals.

A second area of concern is the administration of BST to cows whose milk yield is depressed, because of an undiagnosed disease or a metabolic alteration. In this case BST administration could be an abuse, and adequate veterinary supervision should be available.

Similar considerations arise when BST is administered to animals whose nutritional level is inadequate. The response to BST in this condition will be less, but there are no data clarifying possible changes in the disease incidence. To obtain better milk production, the farmer has a strong economic interest in improving the nutritional status of the animal instead of relying on the abuse of BST administration. Also in this case, however, it all depends on the presence or absence of adequate veterinary control. This must be present not only to supervise BST treatments, but also to safeguard the nutritional and environmental conditions of the animals.

Another possible abuse could be the exploitation of overstretched cull cows (to be eliminated from the farm) in the hope that BST use can extend their milk production. Actually, it is not rational to think that, with such an abuse of BST, sterile or exhausted animals could be restored to profitable lactation.

In conclusion, the fact that treatments every 14 or 28 days with BST require direct veterinary control should not constitute any problem for well-organized farms. In reality, veterinary supervision must already be present for the normal routine of breeding, management and nutrition of the animals.

## FARM NEEDS AND REQUIREMENTS FOR THE CORRECT TREATMENT OF COWS WITH BST

As stated in the last section, good management and proper nutrition are highly important for a correct BST response. If the nutrition is not adequate, the BST response seems to be much reduced. Today, however, no adequate research has been published studying the changes in the incidence of any BST-related disease in undernourished animals. Especially in the first phase of treatment with BST, the animals are in a negative energy balance. This situation can be more dangerous in BST-treated animals that are already undernourished.

This fact concerns not only the abuse or misuse of BST, but also the organizational problems of the farm on which the animals are underfed. Adequate veterinary control is a guarantee not only against undernourishment, but also against abuse of BST. Good buildings and environmental conditions before and during the treatments with BST are of fundamental importance.

Regardless of the local tolerance of BST at the site of the injection, it should be stressed that the amount of material introduced in a single cow is small : e.g., less than 2 ml every 14 days. The injection of such a small amount of material should not cause any marked pain to the animal. Certain sites, however, are less sensitive than others to the injection of exogenous substances. Thus, clear guidelines for injection sites must be drawn up with the purpose of rendering the injection less bothersome for the animals.

The subcutaneous or intramuscular administration of BST causes a mild reaction of the tissues at the site of injection. The resolution time of the tissue reaction is estimated at about 2-4 days for the intramuscular administration, and about 2-4 weeks for the subcutaneous injection. The swellings may be slightly tender to the touch, but they are not easily noted by the stockmen. As for all intramuscular or subcutaneous administrations, the use of adequate hygienic norms helps to prevent abscesses and complications.

There are no available data that demonstrate that the use of BST per se increases the incidence of complications related to the injection procedure. The first weeks of BST treatment, however, promote a negative energy balance. If the nutrition of the animals is poor and the management is bad, it is possible to forecast a certain risk of complications due to the injection procedure.

This event refers to a single defaulting farm rather than to farms in general. Thus, regardless of whether BST is used or not, it would be much better that the animals be guaranteed optimal environmental and nutritional conditions.

BST injection obviously necessitates the handling of the cows. The animals are anxious about restraint and handling rather than about the injection per se. Thus, BST treatments are in practice no problem on adequately equipped farms on which handling is routinely done. Indeed, cows rapidly get used to any routine practice and do not fear it, e.g. having a milking machine put on twice a day!

Thus, if we are talking about suitable farms with adequate handling facilities, the execution of one injection every two weeks or every month does not put the animal under any particular stress. The situation is different if facilities are inadequate and, therefore, forced immobilization for BST injection induces fear and distress.

Other important worries are related to the misuse of BST. There is the possibility that the recommended dose rates could be increased. In this case the milk production does not significantly increase. This misuse is self-eliminating because it becomes very expensive and unproductive and will soon be discontinued. It should be stressed that even when cows receive their total life-time dose of BST in a two-week period, they remain healthy, and subsequent slaughter and examination show no significant abnormalities.

Another very important practical problem in dairy husbandry is that genetic selection of superior cows has inevitably led to selection for higher peak yields. According to some authors, the genetically high-yielding cows are animals that have naturally-occurring high levels of endogenous somatotropin. The use of BST could maintain equivalent levels of somatotropin in animals that are not so genetically selected.

The use of BST could make it less necessary to select cows genetically for higher milk production. Thus, with less pressure to select for milk yield, more emphasis can be placed on resistance to mastitis and diseases, on better temperament, on lower incidence of lameness, and so on.

*Chapter 4*

# INFLUENCE OF COW BREEDS ON RESPONSE TO BST TREATMENTS

The somatotropin level in the blood is related to milk production and so it might be expected that the response to BST would be different in high- and low-yield cows, or among breeds with high or low production potentials. The scanty data available do not demonstrate any higher response to BST in breeds that are potentially lower milk producers. Similarly, within any one breed, it is not clear whether in cows with a lower production potential the effect of BST is greater than in cows that have a higher production potential.

To evaluate the characteristics of dairy cows, the so-called *cow index* is used. It is calculated on the basis of milk production, fat and protein content, the conformation of the mammary glands and the body, genetic characteristics of the parents, etc.. The improvement in milk production caused by BST affects only one of the above parameters and so produces changes of little importance to the cow index.

However, the use of BST may become an abuse if the only parameter adopted for the selection of the sires is the milk production, measured on certain days previously set for this test. The secret administration of BST may then induce an increase in milk production which is not related to the genetic potential of the animal but is only a biological response to the BST administration.

This misuse may cause an alteration of physiological characteristics of the animals and constitute a *swindle* if the sires' price is based only on the relative milk production. In practice, this *swindle* can be eliminated with a compulsory rule to keep a register of the treatments on each animal. Because the treatment is given every two weeks or every month, there does not seem to be any particular difficulty in keeping this register up-to-date in well-organized farms.

It should be noted that selection for higher milk yield has mainly improved the initial yield during the first part of lactation. In fact the genetic variance in this period is larger than for the daily milk yield at later stages of lactation. High-yield cows suffer during the post-partum energy deficit with adverse effects on fertility, occurrence of ketosis and low protein content in milk. Because milk yield and persistency are negatively correlated, the BST treatment will make it possible to attach less importance to the initial milk production and to improve the persistency.

In conclusion, as previously indicated, if selection can be performed with less emphasis on milk yield, more attention can be paid to resistance to mastitis, to better temperament, to lower incidence of fluid filled bursal swellings of carpus and tarsus, etc.

All these aspects of the production benefits on the farms must be considered together with the characteristics of the milk and meat utilized as food. The problem is to determine whether the milk produced by cows on BST treatment has any side-effects for humans. This topic will be discussed in detail in the following chapters.

# 5

# Characteristics of the milk produced by BST-treated cows

Milk is without any doubt an important food which can be consumed directly or can be utilized for the production of many types of food. In the chain between cow and consumer, milk from a large number of cows, from a large number of farms and generally from several regions is transported, bulked together, homogenized, pasteurized or sterilized and then either drunk or further processed. Milk is in fact one of the most widely variable food stuffs that we eat. Moreover, milk has a very complex composition, with specific biophysical characteristics that can easily be modified.

BST administration causes an increase in the production of milk; this raises the question of the milk characteristics after treatment. Milk must undergo many different technical procedures in order to transform it into the different food stuffs. Thus, to evaluate its properties in relation to subsequent processing, it is not enough to determine simply the repartition of its constituents.

In fact, the overall characteristics of milk are made up of the reciprocal combinations and concentrations of its constituents, so that even relatively moderate variations in some of these characteristics can alter the chemical and/or chemicophysical properties of the milk, causing important changes in its suitability for processing.

The characteristic properties of milk can be differentiated into three fundamental categories regarding:
— the *quality*, defined by:
  organoleptical characteristics,

possible microbiological contaminations,
eventual chemical contaminations;
— the *composition*, which influences the characteristics of:
the processing,
the production yield,
the nutritional values of the product;
— the *biophysical properties*, that condition:
the processing suitability,
the fermentation characteristics,
the product yield.

From the point of view of the consumer, it is necessary that the milk has good organoleptical characteristics and a minimal non-pathogenetic microbiological contamination; with no contamination by chemical agents or pathogenic micro-organisms.

It is also important to safeguard the nutritional value of the milk and the productive yield. For example, calcium content and pH value have a marked influence on milk stability and renneting properties. On the other hand, the colloidal properties of the milk must be maintained stable because they condition many processing characteristics. The same consideration can be made regarding the fermentation properties of the milk.

Even if the amount of milk is increased by BST, if all these general properties are not maintained at their normal values, then during manufacturing there will be either a lower yield or a lower product quality.

## INFLUENCE OF BST TREATMENT ON THE CARBOHYDRATE, FAT AND PROTEIN COMPOSITION OF MILK

The daily administration of BST causes no marked variation in the total protein, lactose and fat contents of milk. Obviously, it is important to provide the animals with an adequate food intake in order to avoid a negative energy balance. In this negative condition, the milk content is modified in both fatty acids and proteins, with or without BST treatment of the cow.

The two- or four-week treatments with prolonged-release BST cause a milk production quite similar in protein and fat contents to that of untreated animals, but it is easy to find cyclic variations both in the amount of milk and in its composition.

As previously indicated, the possible increase in lactose output during BST treatment is related to the increased diversion of glucose to the mammary gland. In most tissues the enzyme *galactosyl-trans-*

*ferase* supports the transfer of a D-galactose residue to the N-acetyl-glucosamine, according to the reaction :

UDP-D-galactose + N-acetyl-glucosamine →
$\quad$ → UDP + D-galactosyl-N-acetyl-D-glucosamine $\hspace{2cm}$ (5.1)

In the extra-mammary tissues this reaction is a step in the biosynthesis of the carbohydrate portion of galactose-containing glycoproteins. In the lactating mammary gland, however, D-galactose is a precursor in the synthesis of lactose (a disaccharide of D-galactose and D-glucose).

In fact, galactosyl-transferase is very active with N-acetyl-glucosamine, but only feebly active with D-glucose as galactosyl acceptor. When lactation begins after parturition, in the lactating mammary gland the specificity of galactosyl-transferase changes, now transferring the D-galactosyl group to D-glucose at a very high rate, and making lactose according to the reaction :

$$\text{UDP-D-galactose + D-glucose} \rightarrow \text{UDP + D-lactose} \hspace{1cm} (5.2)$$

The formation of α-*lactalbumin* changes the specificity of galactosyl-transferase. The α-lactalbumin is an enzyme modifier and its synthesis in the mammary gland (regulated by the somatomedin promoting lactation) leads to the formation of an α-*lactalbumin=galactosyl-transferase complex*, i.e., *lactose synthase*. Thus, by hormone action (IGF-I?) the synthesis of lactose is triggered in the mammary gland by the formation of a specificity-modulating subunit of lactose synthetase (Fig. 5.1).

Lactose exists in two anomeric forms designed α and β, that are stereoisomers differing only in the configuration about the carbonyl carbon atom. The properties of the two anomers are different in that the β anomer has a sweeter taste than the α anomer. Moreover, the β is more soluble than the α anomer ; the α anomer may crystallize when ice cream is stored in the freezer for a long time, giving the ice cream a sandy texture.

In humans, before it can be used by the body, ingested lactose must first be enzymatically hydrolyzed in the cells lining the small intestine, to yield its hexose units :

$$\text{lactose + H}_2\text{O} \xrightarrow{\text{lactase}} \text{D-galactose + D-glucose} \hspace{1cm} (5.3)$$

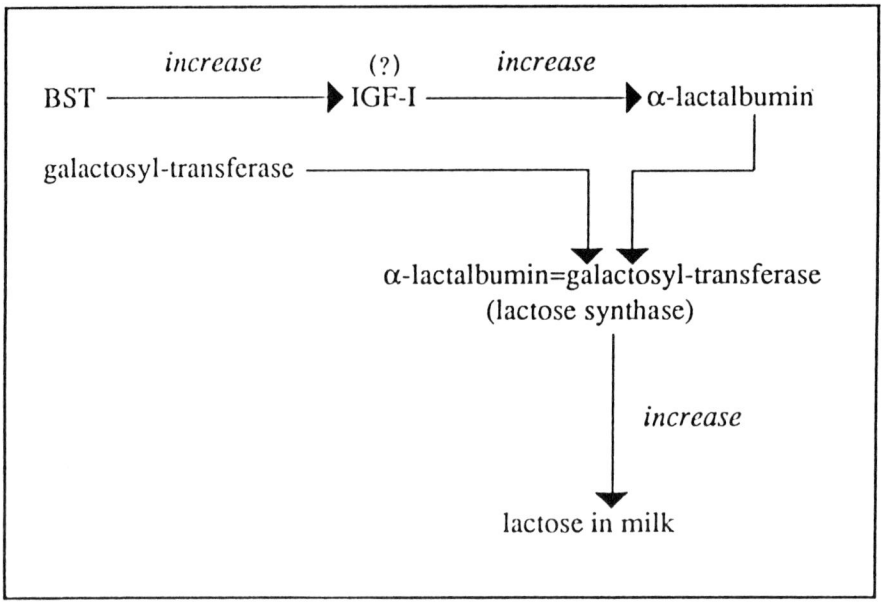

*Fig. 5.1*

In humans, the monosaccharides D-galactose and D-glucose so formed are absorbed into the blood and pass to the liver, where they are phosphorylated and converted into intermediates of the glycolytic sequence.

The *lactose intolerance* of most humans is due to the disappearance or the decrease in the lactase activity of the intestinal cells after childhood. Thus, lactose can no longer be completely digested and adsorbed, and the lactose remaining in the intestinal tract causes discomfort and diarrhea. There is no evident association or relationship between lactose intolerance and galactosemia ; the former is very common, the latter very rare.

D-galactose is first phosphorylated at the expense of ATP by the enzyme *galactokinase* :

$$ATP + D\text{-galactose} \xrightarrow{Mg^{2+}} D\text{-galactose-1-phosphate} + ADP + H^+ \quad (5.4)$$

In the liver, galactose 1-phosphate reacts with UDP-glucose to yield UDP-D-galactose and glucose 1-phosphate, catalyzed by the enzyme *UDP-glucose : α-D-galactose 1-phosphate uridylyltransferase* :

D-galactose 1-phosphate + UDP-D-glucose →
→D-glucose 1-phosphate + UDP-D-galactose   (5.5)

The galactose residue of UDP-D-galactose is then enzymatically epimerized to yield UDP-D-glucose by the enzyme *UDP-glucose 4-epimerase* :

$$\text{UDP-D-galactose} \leftrightarrow \text{UDP-D-glucose}$$

The UDP-D-glucose is then cleaved by *UDP-glucose pyrophosphatase* to D-glucose 1-phosphate, which is converted into glucose 6-phosphate by *phosphoglucomutase*.

UDP-D-glucose + PPi → UTP + D-glucose 1-phosphate      (5.6)
D-glucose 1-phosphate → D-glucose 6-phosphate          (5.7)

This sequence of reactions (5.3 ; 5.4 ; 5.5 ; 5.6 ; 5.7) is not only responsible for the conversion of D-galactose into D-glucose, but is also involved in reverse in the cow mammary gland for the synthesis of D-galactose, required in the formation of milk lactose.

In the genetic deficiency disease *galactosemia*, the enzyme UDP-glucose : α-D-galactose 1-phosphate uridylyltransferase is genetically defective, preventing the overall conversion of D-galactose into D-glucose. Thus, D-galactose and D-galactose 1-phosphate cannot be metabolized and accumulate in the blood and tissues. The liver and other organs become enlarged, vision becomes impaired because of the formation of cataracts, and mental retardation occurs.

A higher content of pyruvate occurs in the milk of BST-treated animals. This fact is of practical interest because in some countries (for example, Germany) milk quality is evaluated partly on the basis of its pyruvate content.

The extra-mammary fat mobilization in the first period after the BST injection causes an increase in the percentage of the long-chain unsaturated fatty acids in the milk fat. This can cause a slight variation in butter characteristics, with a slightly improved spreadability.

The protein content of milk increases slightly during the fourth week after BST injection. Also, in the first week after treatment, the fat content increases in proportion to the increase in milk production, and both decrease subsequently because they are linked to the negative energy balance in the first period after the BST injection.

The proteins are an important group of constituents because they influence the properties of milk and of its processing. In the manufacture of cheese, the casein content is of fundamental importance in the cheese yield. On the other hand, the price of milk is frequently based partially on the total protein content : so, it is very important for the ratio between casein and total protein to remain within normal limits.

Generally, the casein content is not particularly influenced by BST treatment, and neither is the composition of the casein itself. An important characteristic is related to the behaviour of the milk micelles, i.e., the association of amphipathic molecules in water into a structure in which their non-polar parts are in the interior and the polar parts on the exterior are exposed to water. In the micelles there are no changes in the ratio between the different caseins, nor in the ratio between calcium, phosphate and citrate.

Bovine β-casein lacks both cysteine and cysteine residues; it has little tertiary structure, and its native conformation resembles a random coil. Thus, all sections of the protein chain are readily accessible to hydrolysis by pepsin and pancreatic enzymes, resulting in a good digestibility. The constancy of the characteristics and of the protein content of the milk whey determines the processability and the digestibility of the milk. As a matter of fact, there are to date very few published data on the behaviour of individual whey proteins in spite of their importance in milk processability.

## INFLUENCE OF BST TREATMENT ON THE ION, ENZYME, VITAMIN AND HORMONE COMPOSITION OF MILK

Treatment with BST does not markedly modify the osmoregulation of milk because the cation flow from the blood serum to the mammary gland, and from the mammary gland into the milk, is increased to the same degree as the milk production.

In the milk produced by BST-treated cows, many elements (namely, calcium, phosphorus, sodium, iron, copper, magnesium and zinc) show concentrations similar to those in the milk produced by untreated cows. This is true also for total calcium, soluble calcium, ionic calcium and micellar Ca, P, Na, K and Mg. Even in the case of BST overdosage, no particular variations seem to be present in the calcium and phosphorus contents of the milk.

The activities of the enzymes of milk are very important because an increased proteolytic activity can break down the casein prematurely and so reduce the cheese yield. On the other hand, such enzymes can pass into the cheese and therefore affect the ripening process. In the milk produced by BST-treated dairy cows, the activities of aminopeptidase, endopeptidase and phosphatase do not differ particularly from those in the milk of untreated cows. Similarly there is no difference in the total protease, although the level of single protease is unknown. BST treatment does not induce any particular variation in enzymes related to fat metabolism and, in particular, to lipase activity.

There are very few published data on the changes induced by BST treatment on the vitamin contents of milk. From the data available the vitamin A, thiamine, riboflavin, pyridoxine, vitamin $B_{12}$ and panthotenic acid contents seem to be unaltered. There is only a small increase in biotin concentration.

The control of the hormonal content of the milk produced by BST-treated cows is very important. The natural concentration of somatotropin in milk is approximately at the detection level of 0.3-0.5 ng/ml. At the doses normally used in current practice, treatment with BST does not modify the somatotropin level in the milk. Only with very elevated doses is there a certain increase in the concentration of somatotropin in the milk.

This topic will be specifically discussed in Chapter 6 because human somatotropin has the capacity to promote the growth of the skeleton and increase body-weight in young humans. Its deficiency in people results in *dwarfism* but, when it is present in excess, it results in *gigantism* and *acromegaly*. In this case, the growth of the hands, feet, and especially the facial bones is greatly accentuated, resulting in a massive "lantern jaw" and heavy brows.

Also the systemic injection of human somatotropin markedly affects carbohydrate metabolism and the administration of excess hormone causes in humans *pituitary diabetes*, brought about because somatotropin inhibits insulin secretion.

In cows, other hormones stimulated by BST treatment in the adaptation of the neuroendocrine system are the *somatomedins*: IGF-I and IGF-II. BST treatment increases the somatomedin concentration of milk, although it remains within the normal limits observed in the milk of untreated cows. The bovine somatomedins show a notable structural similarity to the human insulin-like growth factors. This important topic will be debated in detail in Chapters 6 and 7.

## CHEMICAL AND CHEMICOPHYSICAL CHARACTERISTICS OF THE MILK PRODUCED BY BST-TREATED COWS

The *freezing point* of milk is not significantly influenced by BST and the same is reported of the *pH value*. The milk produced by animals treated with BST maintains the same organoleptic characteristics even after several days of cold storage.

The *somatic cell count* is very important because in the European Community the control of milk quality includes this test. Treatment with BST does not induce generally marked variations in the number of somatic cells present in the milk, although it is possible to find conflicting data. With higher doses of BST (not usually utilized in practice) there can be a significant increase in the somatic cell numbers.

Another very important chemicophysical parameter is milk *stability*. Test with alcohol indicates that this functional property is unaltered by administration of BST. Another test is the stability to heat, because of its relationship to the urea content which is an important part of the non-protein-nitrogen of milk. The non-protein-nitrogen of milk does not undergo any particular variation or shows only a slight increase after BST treatment. In any case, heat stability is not particularly modified by BST treatment.

## PROCESSING PROPERTIES OF THE MILK PRODUCED BY ANIMALS TREATED WITH BST

The *renneting reaction* (milk curdling power, a fundamental part of the cheese manufacturing process) is very sensitive to variations in some of the properties of milk. Genetic selection has been effected primarily with regard to the level of milk production. Different breeds and even different animals within a breed create very different milk compositions. In any case, the genetic selection of the cow is based on milk yield and the renneting properties of the caseins.

The influence of BST on the production of cheese is evaluated in the coagulation (change in liquid state into clot), acidification, composition, syneresis (contraction of gel, with the expulsion of liquid) and organoleptical properties of the milk. Data related to milk handling after treatment with BST may be influenced by the number of animals utilized in the experiments and by the random selection of animal groups. Such evaluations must be conducted on large numbers of animals and samples must be taken throughout the course of the various periods of lactation.

Some results indicate that the cheese-making characteristics of the milk produced by BST-treated animals do not differ significantly from those of the milk from untreated animals. Some studies, however, carried out on a small number of dairy cows show variations in syneresis and in cheese composition.

# Chapter 5

# GENERAL REMARKS ON MILK CHARACTERISTICS AND PROCESSING PROPERTIES

The examination of various chemical, chemicophysical, organoleptical, and processing properties of the milk produced by the BST-treated cows indicates that the major characteristics are not markedly changed by treatment with BST. Other milk components undergo small variations, some of which are linked with the specific mode of action of somatotropin.

Of these, the first is that, after the BST treatment, it is possible to find somatomedin levels higher than the basal values present in the milk of the same animals before treatment. These values are within the physiological limits generally accepted for the milk of untreated animals. There is, however, a similarity between bovine somatomedins and human insulin-like growth factor I, which raises the question of their possible effect on humans. This topic will be specifically debated in Chapters 6 and 7.

Another point that must be taken into consideration is the possible increase in the number of somatic cells present in the milk, as this is one of the parameters utilized in the European Community for the definition of the quality and characteristics of milk.

Finally, it must be noted that all milk characteristics undergo cyclic variations in function of the time from the BST administration. With or without BST treatment, milk from different herds of cows or even the same herd can vary markedly from day to day and, except at the extremes, no differentiation is made as to their commercial use.

The minor BST-induced fluctuations would probably be totally or partially lost in the overall variation. It should also be noted that milk for the consumer does not reflect the natural variations among cows because the commercial processing of milk standardizes the substrate levels and the type of milk available for the consumer (e.g., skim, fat 1 %, fat 2 %, etc.).

# 6

# Changes induced by BST treatment in the plasma and milk concentrations of somatotropin and somatomedins

As explained in Chapter 5, treatment with BST can induce some changes in the plasma concentration of somatotropin and IGF-I that can lead to modifications in the concentration of those hormones in the milk and, perhaps, in the product derived from it. This question is examined in this Chapter 6.

Moreover, it is known that at the end of their activity as milk producers, cows treated with BST are slaughtered and their meat is utilized for food, either as it is or after industrial processing. This subject is discussed in Chapter 7.

## PHYSIOLOGICAL CHANGES IN BLOOD LEVELS OF SOMATOTROPIN AND IGF-I DURING GROWTH AND LACTATION

To evaluate the potential risk related to the variation of hormone levels induced by BST treatment, it is important to know the normal physiological levels of those hormones (somatotropin and IGF-I) in untreated animals.

Somatotropin levels may be measured by radio-immunoassays (RIA), which reach a sensitivity of 0.3-0.5 ng/ml in blood plasma or skim milk. As a function of low, medium or high concentrations of somatotropin, the coefficient of variation among control samples with known contents (inter-assay coefficient of variation) is good, being about 10 %.

The determination of insulin-like growth factor concentrations (IGF-I) may be also measured by radio-immunoassay, with a coefficient of variation among the control samples (with known contents) of about 12 %.

Up to the age of six months, blood can be collected via permanent cannula inserted in a venous vessel (for example, in vena caudalis). After the age of six months, blood can be collected by extemporaneous insertion of a needle or by a permanent cannula inserted into the jugular vein. The amount of blood needed for the radio-immunoassay is around 10 ml. Due to the chemicophysical characteristics of both somatotropin and IGF-I, blood plasma and whole or skimmed milk can be used for these assays.

During growth (first year of life), the somatotropin concentrations show mean plasma values between 4 and 16 ng/ml in males, and between 3 and 9 ng/ml in females (Brown Swiss breed). During this period of growth the amplitude in the mean plasma levels of somatotropin goes from 4 to 50 ng/ml, with daily individual peak values even higher than 100 ng/ml. The frequency of these episodic daily secretions goes from 2 to 10 in a 24-hour period. During growth, the average plasma concentrations of somatotropin are around 25 ng/ml in bulls, 14 ng/ml in steers and 10 ng/ml in heifers (Fleckvieh breed).

During growth, the blood somatotropin variations are not linked to a variation of its basal secretion, but to an increase in amplitude of secretion levels. The plasma level of somatotropin decreases more rapidly in females than in bulls which, independently of daily or monthly variations, show higher values.

During growth (first year of life) the mean blood concentration of IGF-I is from 400 to 1,800 ng/ml in males, and from 300 to 900 ng/ml in females (Brown Swiss breed). The main tendency to an increase in IGF-I values occurs around puberty (particularly in males). The average concentrations of IGF-I in blood are around 1,400 ng/ml in bulls, around 1,300 ng/ml in steers and around 1,100 ng/ml in heifers (Fleckvieh breed). Therefore, the differences between the blood concentration of IGF-I in bulls and heifers (+ 30 %) are markedly less than the differences between the blood concentrations of somatotropin in bulls and heifers (+ 150 %).

During lactation, the mean blood concentration of somatotropin (Brown Swiss breed) increases in the first two months after parturition (for example, from about 7 ng/ml to 25 ng/ml, with a milk production of about 28 kg/day), decreases 4 months after parturition (for example, about 10-11 ng/ml, with a milk production of about 24 kg/day) and remains at the basal values for the last 5 months of the lactation (for example, with a milk production of about 10 kg/day at the 44th week). The post-partum increase in mean blood concentration of somatotropin is related to an increase in episodic secretory activity, with increase in both basal concentrations and amplitudes, the peak levels of amplitudes being below 30-40 ng/ml.

After the period of peak lactation, the decrease in mean blood level of somatotropin is due to a decrease in both basal value and maximum of amplitudes, there being no modification in the frequency of amplitudes. The data relating to the Brown Swiss breed are similar to that obtained from other breeds, for example, Holstein Friesian.

## CHANGES BY BST IN COW BLOOD LEVELS OF SOMATOTROPIN AND IGF-I

Treatment of cows with BST induces significant variations in their blood concentrations of somatotropin, and is affected by the dosage, the type of treatment (daily injections or prolonged-release preparations), the time-course of evaluation (short-term or long-term treatments), the time of blood collection after the BST injection, etc..

The blood concentrations of somatotropin are significantly greater with prolonged-release BST preparations, especially in the first week. For example, 3 days after the subcutaneous injection of 640 mg of prolonged-release preparation of BST, the mean blood concentration of somatotropin increases 3-5 fold (from about 8-10 ng/ml to about 35-45 ng/ml : Fleckvieh breed) and decreases over 8-10 days to mean values slightly higher than the norm (12-14 ng/ml), with a high variability between cows.

Also the plasma concentrations of IGF-I are significantly increased by treatment with BST, especially during the first weeks after the injection. For example, 6 days after the subcutaneous injection of 640 mg of prolonged-release preparation of BST, the mean blood levels of IGF-I increase about 2-5 fold (from 250-500 ng/ml up to above 1,200 ng/ml), with persistence of high values after one week (above 700 ng/ml) and a decrease to the initial values in about two weeks (Fleckvieh breed). In this breed, heifers in the growing phase show a mean blood concentration of IGF-I around 1,100 ng/ml (section 6.1) ;

therefore, the BST induced changes are within the range physiologically observed in growing animals.

## CHANGES IN THE SOMATOTROPIN AND IGF-I CONCENTRATIONS IN THE MILK FROM BST-TREATED COWS

Milk is composed of substrates or precursors transported in the blood and, therefore, circulating hormones such as somatotropin or somatomedins (see also section 8.3) may be transported into the milk by passive or active mechanisms.

*Somatotropin* secretion in the bovine milk is very low because the concentration in skim milk does not reach the sensitivity limit of 0.5 ng/ml. BST administration (for example, 500-600 mg/14 days or 640 mg/28 days) does not seem to increase these values, which remain under the minimal concentration levels. It should be noted that BST is parenterally inactive in humans and may be orally inactivated (see section 7.6).

The concentrations of *IGF-I* in human breast milk are around 30 ng/ml in the 6-8 days pre-partum, decreases to about 14-18 ng/ml in the post-partum days, and stabilizes over the following week at 7 to 8 ng/ml. IGF-I concentrations in human milk at six to eight weeks post-partum range between 13 and 40 ng/ml (with a mean of about 20 ng/ml), and are two- to three-fold higher at six to eight weeks post-partum than at three to seven days post-partum. The plasma levels of IGF-I are highest in young persons, with concentrations reaching two- to three-fold those in adults, with positive correlation between circulating levels of IGF-I and adolescent skeletal growth.

Samples of whole milk obtained from more than 400 untreated cows from many commercial dairy herds have IGF-I levels from undetectable to about 30 ng/ml, with a mean IGF-I concentration of about 2.5 ng/ml. The mean concentration of IGF-I from bulk tank samples from untreated cows of 100 commercial farms is around 5 ng/ml.

There is little published information on the effects of prolonged-release preparation of BST treatment on the IGF-I concentrations in milk. One report shows a two-fold increase in whole milk, although the absolute increase is only in the order of 2 ng/ml. Other data indicate skim milk levels to increase from 26-33 ng/ml to 32-44 ng/ml (Fleckvieh breed) although the difference is not statistically significant.

Overall, treatment with suggested dosages of BST (section 3.1) seems to cause a slight increase in the levels of IGF-I in bovine milk, but the absolute values of concentrations remain within the range found in the milk of untreated cows and are no greater than those found in human milk.

In the case of BST overdosage (administered either in bad or in good faith) the milk yield does not increase even more, but the plasma levels of IGF-I exceed the normal values and this can have repercussions on the milk. For example, subcutaneous administration of a very high dose of BST (2 g) induces an increment of 5-7 fold in the plasma values of IGF-I (from less than 200 ng/ml to a peak of about 1,500 ng/ml) : in this case the concentration of IGF-I in the skim milk of BST-treated cows is 1.5-1.8 fold higher than that found in untreated animals.

Overdosage of BST should be made impossible for the defence of the health and welfare of the animal. In any case, the cost/benefit ratio for the farmer suggests that in bad faith this overdosage is unlikely and unprofitable.

The amino acid sequences of bovine and human IGF-I are identical and the bovine IGF-I could be biologically active in humans if administered parenterally. Although BST slightly increases the milk IGF-I concentrations, they remain within the range found in the milk of untreated cows. The presence in the milk of such levels of IGF-I, or of its fragments, does not seem to constitute any particularly severe problem in humans orally consuming milk because :

(a) the concentrations are very small and within the range detectable in the milk of untreated cows ;

(b) infants are exposed to the same or higher levels of IGF-I in human breast milk ;

(c) such peptides would be readily digested to small peptides and amino acids in the human intestinal tract, and the actual exposure to IGF-I from milk consumption will be much less because of the digestion of IGF-I in the gastrointestinal tract ;

(d) effects of IGF-I administered to humans have only been noted when it has been administered parenterally in large doses ;

(e) pasteurization, sterilization or processing of milk to yogurt or cheese (also for infant formulas) will largely break down IGF-I.

Insulin-like growth factor-II or IGF-II (see section 8.3) belongs to the group of somatomedins and is another growth factor regulated by BST. The factor is less effective than IGF-I and acts by specific receptor interactions. The plasma levels of IGF-II in adult human subjects are approximately 650 ng/ml. The age-dependent pattern for circulating IGF-II levels is different from that of IGF-I : as discussed in section 8.3, the IGF-II levels at birth are low, but reach almost the normal adult levels by one year of age. It should be stressed that, in contrast to IGF-I, circulating levels of IGF-II are not correlated to adolescent skeletal growth. With respect to the biological action of IGF-II, the above discussion on IGF-I adequately covers any concerns for both the insulin-like growth factors.

In considering the human food safety of any particular reported milk level of BST, IGF, etc., it should be noted that levels in untreated cows milk vary widely because cows are the most heterogeneous of farm animals. Their genetic potential and management vary markedly in even one part of a country.

Although milk is perceived by so many as a constant chemically defined solution of healthy ingredients, it is in fact the food stuff that varies most widely in both minor and major constituents. However, in the chain between cow and consumer, milk from a large number of cows, from a large number of farms and generally from several regions is transported, bulked together, homogenized, pasteurized or sterilized, standardized and then either drunk or further processed. The processes of pasteurization, sterilization, processing, cooking, digestion, etc. will denature proteins such as BST, IGF-I, etc.. See also sections 7.4, 7.5 and 7.6.

## POTENTIAL INFECTIVE OR IMMUNOGENIC RISKS TO HUMANS FROM MILK FROM BST-TREATED COWS

It could be observed that a possible stressing effect of BST treatment in cows may induce immunosuppression with the activation of latent viruses. Even if viruses are species-specific, some of them may have a bearing on human risks of immunosuppression.

Viruses present in milk would be, however, eliminated during the pasteurization process and the transmission of bovine viruses through milk could not be a problem of safety for humans in countries using the milk pasteurization. On the other hand, the infection of dairy cows with a tumor-causing virus, namely the bovine leukemia virus, is relatively widespread but, fortunately, the quoted virus seems never to have infected humans, even people drinking raw milk from known infected herds.

The dairy cows may be infected with other viruses (e.g., the bovine immunodeficiency virus) whose effects are not well-characterized. Today, there is no clear evidence that this kind of virus has negative effects in infected cows, but both their influence and the possible BST activation of latent viruses in cows are research fields of great interest.

A possible stressing effect of BST treatment in cows could increase the incidence of typical diseases of dairy cows (e.g., mastitis, fluid filled bursal swelling, etc.) and the consequent use of antibiotics to combact these diseases. With some exception, the incidence of the typical diseases of dairy cows seems to be the same as that in untreated animals. In any case, also to prevent the risk of inducing increased antibiotic resistance in humans, the general regulatory systems currently employed to monitor milk for drug residues should prevent milk contamination and eliminate adulterated sources from commercial sale.

The possible immunogenic effects in humans of the development of circulating IgG antibodies after injection of human somatotropin and/or recombinant derivatives do not seem to be due to the substances per se, but rather to the high level of bacterially derived contaminants in the preparations. The therapeutic use of injections with increasingly purified preparations of human somatotropin leads to a reduction in the number of human subjects developing circulating immune responses, and these do not seem to have clinical implications because there is scanty or no correlation between the immune response and growth rate or final height of the human subjects.

The ingestion of BST protein in cow milk could cause a specific allergic response, even if many people are continuously exposed to a wide variety of proteins with ingestion of any food stuff. Generally, the uptake of intact proteins in adults and neonates does not cause particular problems, but certain genetic or disease states increase the uptake of intact proteins causing an allergic response. Today, there is no evidence that BST is more allergenic than other milk proteins. Moreover, BST represents an extremely minor component of milk proteins (less than 0.000001 mg/ml of milk) in comparison with the concentration of the major milk proteins : 25-30 mg/ml for caseins, 3-8 mg/ml for whey proteins, 0.5-1 mg/ml for immunoglobulins, etc..

Another potential and important problem is the contamination of milk with pesticides and the possibility that a lipolytic effect of BST treatment in cows will mobilize various pesticides from adipose tissue and increase their milk levels. To date, milk contamination with pesticides seems to involve isolated incidences on farms and seems to be unrelated to the general contamination of the cow adipose tissue.

The initial mobilization and subsequent replenishment of fat in adipose tissue occur naturally during the normal lactation cycle in untreated dairy cows. An enhanced lipid mobilization will occur only when increased milk production causes BST-treated cows to be in a negative energy balance, as described in sections 1.3 and 2.1.

In fact, BST treatment of lactating cows immediately induces an increased milk yield, whereas the voluntary feed intake increases after 4 to 6 weeks of BST administration to support increased milk synthesis. As BST does not markedly alter milk synthesis and composition, it should mobilize similar amounts of fat from adipose tissue over a lactation cycle in comparison with untreated dairy cows producing the same yield of milk.

Fat content in fresh milk is the most variable of its major components and is dependent upon many factors, namely diet, breed, stage of lactation, age of cows and season. Changes in the untreated cow's diet can manipulate milk fat content by changing the ratio of grain to forage, by altering the physical characteristics of the forage, by adding concentrates, etc.. These changes modify rumen fermentation and its end-products available for milk fat synthesis.

Different breeds of cattle have wide differences in milk fat concentration. Milk from Holsteins has the lowest fat content (about 3-4 %), whereas milk from Jerseys has the highest fat concentration (about 5-6 %). Milk of colostrum has a relatively high fat concentration (greater than 5 %) which then decreases until 5 to 10 weeks post-partum (about 3-4 %), at which time fat content increases until the end of lactation (approximately 44 weeks) to about 4 % and is inversely correlated to daily milk yield.

As regards the species of fatty acids involved, the mobilization of fat from adipose tissue increases the proportion of long chain (16 and 18 carbons in length) versus short chain (4 and 16 carbons) fatty acids in milk fat. Moreover, in fresh milk of untreated cows the content of fat and its ratio of fatty acids are highly variable throughout the lactation and dependent upon many other factors, including the age of cows and the seasonal changes.

As stated before, it should be remembered that milk for human consumption does not reflect the natural variation in fat content among untreated cows because the commercial processing of milk standardizes the fat levels and leaves the choice of milk fat content to the consumer.

# 7

# Characteristics of the meat of BST-treated bovines

As mentioned in Chapter 6, cows treated with BST to enhance milk production are liable to be slaughtered and their meat is liable to be used directly or indirectly as food. In general, the result of BST treatment is a reduction of carcass fatness and an increase in lean meat.

## EFFECT OF BST ON THE CHEMICOPHYSICAL CHARACTERISTICS OF THE MEAT

Long-term treatments with somatotropin (extracted or recombinants ; for example, 50-200 mg/kg for 9-20 weeks) in various bovine breeds (for example, Belgian heifers, Fleckvieh veal calves, dairy-type heifers, beef steers) do not induce in the muscular tissues (for example, trapezius, semitendinosus, longissimus dorsi, iliospinal, etc.) any gross variation in the dry matter, which maintains its normal percent value (about 25-29 %).

The percentage fat content of the muscles tends to decrease, remaining however in the minimal range (1.4-1.5 %) that renders the meat appetizing from an organoleptic point of view. Furthermore, the percentage of both collagen and juice extraction tends to increase slightly (for example, from 2.25 to 2.50 % for the former, and 24.5 to 28.5 % for the latter). Also the percentage of weight-loss on cooking increases (for example, from 17 to 18 %). In contrast, the pH, the pigmentation and the colour of the meat are unchanged.

The meat quality is evaluated on compressibility, shear force, tenderness, colour, taste, juiciness and overall acceptability. Treatment with somatotropin does not markedly modify most of those characteristics, but slightly decreases have been seen in both the tenderness and the overall acceptability.

The number of the studies carried out and the number of animals treated are relatively small and the overall duration of the treatment (4-5 months) is relatively short in comparison with the possible prolonged use in dairy cows.

In conclusion, the treatment with BST does not seem to modify in any substantial way the chemicophysical and organoleptic characteristics of bovine meat, but more data are required on the possible slight modifications of meat quality which are in the adverse direction, namely collagen content, juice extraction, and weight-loss during cooking.

## RESIDUES OF BST PRESENT IN MEAT AFTER TREATMENT WITH PROLONGED-RELEASE BST

The problem of BST residues in the tissues of treated dairy cows requires some consideration. First, it should be stressed that at this time there is no known technology that can assay BST levels in muscle or other body tissues (other than blood and milk). This is because of the impossibility of extracting and separating the 191 amino acid chain, intact with its structural relationship unaltered, from the other tissue proteins.

The concept of labelling or tagging the BST molecule has been discussed many times but unfortunately it is not a practical proposition. This is because both naturally-occurring pituitary and recombinant BST have in the body very short half-lives (time required for the disappearance or decay of one-half of the somatotropin). Thus the small quantity of the BST molecule is rapidly broken down in the body and its amino acids utilized, for example, to make other body proteins.

In the absence of empirical observation it is possible to make a very simple arithmetical inference. Let us assume that the BST absorbed every day from the injection site requires as much as 24 hrs (!) to be completely broken down in the animal body. If in a 340 kg steer there are both a uniform release of somatotropin from the injection site and a uniform tissue distribution of prolonged-release preparation of BST after a subcutaneous treatment at 640 mg/28 days, it would have a mean concentration of about 0.022 g of BST/day/340 kg of steer = 0.000000065 g of BST/day/g of tissue = 65 ng of BST/g of tissue (e.g., blood plasma)/day.

As a matter of fact, this theoretical tissue concentration is not in practice achievable because the BST injected undergoes quick tissue metabolism and inactivation. In reality, in this condition one sees : (a) a mean actual value of about 30 ng of BST/g of blood plasma, 1 day after the beginning of the treatment ; (b) a mean actual peak of about 45 ng of BST/g of blood plasma, 3 days after the injection ; (c) a subsequent decrease to the normal blood values in 8-10 days (10-12 ng/g blood plasma). Thus the theoretical value of 65 ng of BST/g is not achieved in the blood plasma even in the peak period.

This is because the BST tissue concentration is determined both by BST distribution in the tissues and by its metabolization which rapidly degrades and inactivates the molecule. For example, in lactating Holstein cows, the decline in blood BST concentration through time follows a biexponential curve : the initial distribution half-life is about 7 minutes, while the terminal distribution half-life is about 30 minutes.

The actual quick distribution and metabolization require a continuous release of somatotropin from the BST injection site to maintain the biological activity.

Let us assume that the subcutaneous treatment with BST prolonged-release preparation at the dose of 640 mg/28 days is performed in a 560 kg cow. In this case, it would have a theoretical mean tissue concentration of about 0.022 g of BST/day/560 kg of cow = 40 ng of BST/g tissue/day. Furthermore, if BST is concentrated only in the cow's edible tissues, the theoretical concentration in the meat will increase from 40 ng of BST/g to about 100 ng of BST/g of meat, equal to 100,000 ng of BST/kg of meat, corresponding to a 0.1 mg of BST/kg meat/day.

Now we can consider an unrealistic hypothesis in which the BST : (a) has the same biological activity as the human one ; (b) is active by the oral route in man ; (c) is not broken down in the human gut. It is well known that to obtain an actual biological activity in human subjects who have a somatotropin deficiency (and therefore are very sensitive to the somatotropin effect) it is necessary to administer from 2 to 4 mg by the intramuscular route, three times a week, for a period of 1 or 2 years. Considering that the theoretical level of BST in the cow's meat in the above--mentioned example is 0.1 mg of BST/kg meat/day, it is possible to conclude that, to reach an active dosage, a boy should eat from 20 to 40 kg of meat per day, three times a week, for a period of 1 or 2 years.

## THE FUNDAMENTAL DIFFERENCES BETWEEN HUMAN AND BOVINE SOMATOTROPIN

Contrary to what is hypothetically indicated in section 7.2, human somatotropin and bovine somatotropin are very different from each other.

One of the differences concerns the *amino- and carboxyl-terminal residue* that in human somatotropin (hST) is phenyl-alanine, while bovine somatotropin contains amino-terminal alanine and phenyl-alanine in approximately equal amounts.

A second difference concerns the *isoelectric point* that is 4.9 for hST and 6.8 for BST. A third difference between human and bovine somatotropins is related to the *molecular weight*, as indicated hereafter according to various methods of evaluation (Table 7.1).

Table 7.1.

| Type of somatotropin | Common literature values | Values estimated by column | | |
| --- | --- | --- | --- | --- |
| | | Sephadex G-100; pH 7,5 | Sephadex G-100; pH 6,0 | Polyacrylamide disc gel |
| human | 28,000 | 20,500 | 21,000 | 23,000 |
| bovine | 45,000 | 22,000 | 28,000 | 27,000 |

A fourth difference concerns the *amino acid composition* of which the number is highly different for most single amino acids in bovine somatotropin with respect to the human one, as shown in Table 7.2.

## POSSIBLE BIOLOGICAL ACTIVITY OF BST IN HUMANS

The significant differences indicated in section 7.3 between human and bovine somatotropins make clear the species-specificity, which is demonstrated by the lack of effect when BST is administered parenterally to humans. As a matter of fact, with the exception of monkey somatotropin, other species' somatotropins are inactive in humans. Only monkey and human somatotropins are active when injected parenterally in a human pituitary dwarf. Monkey somatotropin induces marked physiological effects when injected in rhesus monkeys, whereas pituitary bovine somatotropin has no effect.

Pituitary human somatotropin is effective in humans, while somatotropins derived from bovine, ovine, whale and porcine pituitaries are ineffective. The ineffectiveness seems to be related to a form of specificity in the somatotropin molecules due to differences in chemical configuration. As previously shown, although pituitary bovine somatotropin and pituitary human somatotropin both have 191 amino acids, there is approximately a 35 % difference in their amino acid sequences. Consequently, the pituitary bovine somatotropin does not compete with human one for binding sites in human cell membranes in vitro and pituitary bovine hormone does not bind to somatotropin receptors in human tissues.

Table 7.2

| Amino acids tested | Human somatotropin | Bovine somatotropin |
|---|---|---|
| Lysine | 9 | 12 |
| Histidine | 3 | 3 |
| Arginine | 11 | 13 |
| Aspartic acid | 20 | 16 |
| Threonine | 10 | 12 |
| Serine | 18 | 12 |
| Glutamic acid | 26 | 25 |
| Proline | 8 | 6 |
| Glycine | 8 | 10 |
| Alanine | 7 | 13 |
| Half cystine | 4 | 4 |
| Valine | 7 | 6 |
| Methionine | 3 | 4 |
| Iso-leucine | 8 | 7 |
| Leucine | 26 | 24 |
| Thyrosine | 8 | 6 |
| Phenyl-alanine | 13 | 13 |
| Tryptophane | 1 | 1 |

The specificity of the human somatotropin receptor for human somatotropin and the ineffectiveness of somatotropin from non-primate species in humans lead to the term *species-specific* being applied to the somatotropins. As a matter of fact, this terminology is not technically correct, because BST is more or less effective in rats, goats, pigs, sheep, etc.. However, the terminology has continued with the understanding that it indicates a different sensitivity as one goes up the phylogenetic tree, with humans and monkeys being unaffected by somatotropins from lower species.

The ingestion of bovine milk or meat by humans leads to the digestion of the BST molecule in the gastro-intestinal tract, where proteolytic enzymes break down the somatotropin into small peptides and amino acids. A doubt could arise regarding the levels of chymotryptic digestion of somatotropin, because a limited digestion of somatotropin (for example, only 25 %) does not modify the biological activity which remains unaltered. A more extensive or total chymotryptic digestion of somatotropin leads to the total loss of biological activities.

## THE BST "CHYMOTRYPSIN CORE"

Limited tryptic digestion of BST retains some of the activity of intact BST when injected into hypophysectomized rats with :
(a) a progressive loss of growth-promoting activity with the increase in the number of bonds broken between the amino acids ;
(b) a marked decrease in activity when the number of bonds split is greater than three.
For example, by limited tryptic digestion it is possible to obtain two fragments of BST, one of which retains approximately 1 % of the activity of intact BST and one of which is inactive ; the two fragment recombination results in 10 % of the activity of native BST.
Clinical studies in humans do not seem to substantiate that injected BST fragments are biologically active. Only very limited effects are present in patients injected with large doses (5 to 100 mg/day) of the BST tryptic digests. Moreover, the high variability in the effects makes them difficult to ascribe to the treatment.
With a particular method of fractionation, a BST of 21,500 kD may be separated into two peptides with molecular weights of 5,000 and 16,500 kD, respectively. The lighter peptide is biologically more active, even though each of the two peptides is less active than the parent hormone.
The observation that the bovine growth hormone can survive a significant degree of digestion by chymotrypsin not only suggests that activity does not require the entire molecule, but also leads to the speculation that the digest (the *chymotrypsin core*) might be active in man.
This idea is the most promising yet advanced in support of the hypothesis that a fairly simple moiety of an animal growth hormone might be found that is active in human subjects. However, the specific activity of the presumed active fragment is very low, and large doses are required for an unequivocal effect to be seen. A great deal more evidence

must be obtained, especially on the response of human subjects, before it will be possible to say that the goal has been attained.

## RELATIONSHIP BETWEEN BST AND HUMAN DIGESTIVE TRACT

BST is a protein and, after the ingestion, it would be degraded in the gastrointestinal tract much as any other protein whose digestion products enter the blood almost entirely as free amino acids. Although the transport of intact proteins from the intestine to the blood has not been markedly investigated in adult animals, it is known that proteins may be absorbed intact in healthy humans. This event could start off a chain of steps resulting in circulating antibodies to food proteins. However, in general no adverse reactions occur in humans in response to the absorption of the intact proteins.

Injected BST is biologically active in rats and, therefore, this species is suitable as the model to investigate whether the protein would be active when ingested. However, in such studies, no physiological effects are evident after the oral administration of large doses of BST.

A question can be raised concerning the absorption of BST or BST fragments in the digestive tract of newborn infants prior to closure time, or in humans with impaired protein digestion. The gut of the newly born infant is permeable to foreign proteins, as evidenced by the appearance of specific antibodies against proteins; however, the protein absorption seems to be no more significant in fullterm neonates as compared to adults because the amount of protein absorbed is in the order of 1:50,000 of the amount of protein ingested. Even in preterm infants or neonates a complement of enzymes for the protein digestion is present, although it is limited.

The closure time of gut permeability to proteins in the newborn human may be as long as 3 months after birth or, in other studies, occurs within the first 30 hours or even before birth. Even though proteins are non-selectively taken up into gastrointestinal tract of neonates, their transport into the peripheral circulation does not necessarily follow uptake.

Although the *chymotrypsin core* hypothesis could be true, it does not seem to have marked practical significance with regard to possible harm from the oral ingestion of BST fragments produced by gastrointestinal digestion for many reasons.

To obtain the limited-activity BST fragments in the human gastrointestinal tract, it would be necessary to have very mild biochemical conditions. However, the breakdown of proteins seems to be generally quite extensive. At the same time, the concentration of BST in milk and/or meat, from which such fragments could be derived, is extremely small in comparison with the dose necessary to have biological activity by injection.

Although there is a pituitary somatotropin fragment (bGH 96-133) which is biologically active in humans, this peptide is broken down in the gut of the human much more rapidly than the parent somatotropin. Even if it were not digested, to have a biological effect, the amount of cows milk or meat that would be needed to be ingested daily is not compatible with the ability of human subjects (see section 7.2).

The issues are essentially the same for recombinant bovine somatotropin. Since its amino acid composition and sequence are essentially homologous to pituitary somatotropin, and previous studies with recombinant bovine somatotropin have not demonstrated any significantly different biological responses, it should not be considered differently from pituitary somatotropin. The same considerations apply to peptides derived from recombinant material because chemically they are the same as peptides derived from pituitary bovine somatotropin.

In conclusion, BST in the meat would be largely or entirely digested in the gut lumen of humans, particularly in nanogram amounts. Even if some small amounts of intact BST could be absorbed into the human blood stream, it does not seem likely that it could produce significant effects in humans for two reasons. In the first instance somatotropins are species-specific and the bovine somatotropin seems to have no effect in humans. Additionally, small amounts of somatotropin, even if absorbed intact after oral ingestion, do not seem likely to produce any effects.

As to the possibility that fragments derived from BST may have an effect in humans, this, too, seems most unlikely. For example, one biologically active peptide (i.e. bGH 96-133) is derived from bovine pituitary somatotropin and does have activity in humans. However, if it appears in the gut, this peptide is more rapidly degraded than intact growth hormone itself. In addition, when this material is administered parenterally it is necessary to give a large amount to demonstrate any effect in hypopituitary humans.

# 8

# Biological and pharmacological manipulations alternative to the use of BST

The modification of the biological characteristics of the animal (for example, to increase the production of milk or to change the meat characteristics) derives from both economic needs and consumer requirements. For example, the wide publicity of cardiovascular damage due to fatty foods creates a consumer requirement for leaner meat. As a consequence, the farmers have tried to meet these requirements.

## BIOLOGICAL MANIPULATIONS AND THE PRACTICAL USE OF BIOTECHNOLOGICAL PRODUCTS

The economic desire to obtain, in a relatively short period of time, a quantitative variation (for example, an increase in milk production) or a qualitative variation (for example, a reduction of fat in the meat) creates the biological need to modify morphofunctional characteristics of animals which have developed over a long period of time. Progress in biotechnology makes it possible to modify those characteristics which remain less profitable, or unpleasant, or dangerous (for example, the significant presence of fat in the meat).

The practical contradiction becomes by the fact that, once a biotechnological way to make these modifications is known, people tend aprioristically to refuse, because it alters the *natural* functions of the animal. This takes no account of the fact that since man first domesticated animals he has been breeding them and selecting them to change their *natural* characteristics to those that he desires or that make the end product more useful or acceptable.

It is unclear why the biological *natural* characteristics of the animal must be retained as aprioristically optimal. They have, after all, been derived from a long adaptation process to exogenous events, such as environmental conditions, breeding systems, kind of feeding, frequency of disease, etc.. The changed *natural* characteristics of the animals may be in a positive direction. For example, the human animal when it was exposed to cold winters, to food deficiency, to disease, etc., had natural morphofunctional characteristics different from those of humans that eat a balanced diet, are sheltered from cold winters by central heating and are protected from disease by antibiotics, vitamins, hormones, etc..

A second comment concerns the *semantic* aspect that influences exogenous treatments of, for example, the biotechnological type. If substances are listed as vitamins, or salts, or hormones, etc., there is a favourable acceptance of vitamins and salts, and an inevitable refusal of hormones. Thus hormones, which cannot be dangerous and may even be useful, are a priori rejected while vitamins, which can be dangerous and non-useful, are a priori accepted.

Thus, every substance needs to be discussed singly and scientifically, according to its documented biodynamics, biokinetics, biometabolism, etc., keeping in mind that the variation of only one of the parameters of use (for example, the dose) determines the appearance of useful or dangerous effects.

## ALTERNATIVE MANIPULATIONS BY TREATMENTS MODIFYING SOMATOTROPIN LEVELS

The relationship between the levels of somatotropin and the milk production potential means that manipulations, as alternatives to the BST, tend to raise the levels of somatotropin :
— *increasing somatotropin endogenous release* : GRF, peptides, amino acids, adrenergic agents ;
— *inhibiting the secretion of the somatostatin,* which inhibits the release of somatotropin.

In the first group of substances the *GRF* or *growth-hormone releasing factor* has an important role in increasing the milk production in lactating animals by an action on the mammary gland, which is probably mediated by the IGF-I. There are also some cyclic analogues that are more stable than the GRF itself.

*Alpha-adrenergic* agents, such as *clonidine*, stimulate the release of somatotropin, probably by a direct action on the pituitary cells. It should be stressed that clonidine reduces the mean arterial blood pressure because of the interaction with both central and peripheral adrenergic transmission. In fact, clonidine interferes with the central pre-synaptic $\alpha_2$-adrenergic receptors, inhibiting the release of nor-adrenaline, reducing the activity of the sympathetic nervous system. At the peripheral level, clonidine blocks the post-synaptic $\alpha_1$-adrenergic receptors, with a consequent reduction of vascular peripheral resistances.

Among the *peptides* stimulating somatotropin secretion which stimulates milk production, the analogues of met-enkephalin have a certain importance, and in particular a hexapeptide that does not seem to act with the same mechanism as GRF. Met-enkephalin belongs to the group of endorphins, the *body's own opiates*. Enkephalins formed in the CNS bind to specific receptors in specific neuronal cells and induce analgesia, deadening of pain sensation. Enkephalins become bound to cerebral receptors also binding to opiate drugs (e.g., morphine, heroin, etc.). The first two natural opiate-peptides identified and isolated from the brain are two pentapeptides that differ only in the terminal amino acids : met-enkephalin and leu-enkephalin, according to the amino acid in position 5 of the sequences that could be methionine or leucine, respectively.

The amino acid *arginine* (that presents as side-chain a positively charged guanidinic group) stimulates the secretion of hormones, like insulin and prolactin. In particular, its intravenous infusion in lactating dairy cows increases the somatotropin circulating levels, without reaching values able to increase the production of milk. In contrast, in pregnant cows the intravenous infusion of high doses of arginine (0.1-0.5 g/kg/day) raises not only the somatotropin levels but also the milk production.

As indicated in Chapter 7, some *somatotropin fragments* have one or more characteristic functions of the parent molecules, but do not reproduce completely the biological action that is at the basis of increased milk production.

## ALTERNATIVE MANIPULATIONS BY SOMATOMEDINS (IGF-I AND IGF-II)

Hormonal regulation of lactation is extremely complex involving many different hormones, nutrient availability and multiple interactions between the endocrine system and genetic and environmental factors. Somatotropin and its related growth factors (insulin-like growth factors ; or somatomedins ; or IGF-I and IGF-II) are believed to play an essential role in this biochemical field.

Pituitary and recombinant-derived bovine somatotropins increase milk production, probably by the IGF-I intervention. Unfortunately, conclusive *in vivo* studies defining the specific molecular role of IGF-I in lactating mammary tissue are not yet available and contradictory results have been published using *in vitro* approaches. Clearly, additional work is required to reconcile these differences to ascertain the molecular functional role of IGF-I in mammary metabolism during lactation.

Somatomedins are present in high concentration in mammary secretion during pregnancy and in colostrum ; they are also present in milk, but at much lower concentrations. The decline in concentration of milk insulin, IGF-I and IGF-II during the early post-partum period is not due to a dilution effect as milk yield increases, because the total amount of IGF-I to enter bovine milk diminishes during the first five weeks of lactation. This may be associated with a decline in receptor numbers, although past studies indicate an increase in receptors for IGF-I and IGF-II in the immediate post-partum period.

In dairy cows, there are more receptors for IGF-II than for IGF-I in late pregnancy and early lactation : this difference in receptor population is reflected in the concentration of these hormones in milk. However, in mid-lactation, bovine mammary tissue contains equivalent populations of type I and II somatomedin receptors and again this is reflected in the milk concentration of these hormones. In the cow there appears to be a single population of type I somatomedin receptors, in contrast to the two insulin receptor populations.

A truncated form of IGF-I is present in colostrum, where it occurs in high concentration. This modified IGF-I (lacking the N-terminal tripeptide Gly-Pro-Glu) possesses enhanced biological activity compared with native IGF-I. This truncated form is approximately 4-times more potent than native IGF-I in stimulating DNA synthesis in bovine mammary tissue in collagen gels. This difference in potency is not due to a greater affinity of the IGF-I receptor for the IGF-I truncated form.

Low-molecular-weight binding-proteins are secreted by bovine mammary tissue and may require the N-terminal tripeptide (missing in the truncated form) to bind IGF-I. Although not yet amply demonstrated, such a difference in the ability of binding proteins to recognize the truncated molecule may result in greater availability of this molecule for the receptor on the mammary epithelial cell, and hence its greater biological activity.

The uptake of IGF-I by the mammary gland appears to decline with the onset of lactation, as indicated by falling concentrations in milk and reduced total IGF-I in milk. The high concentration of IGF-I in mammary secretion during pregnancy and its potency in inducing cell division in mammary epithelial cells suggest that this somatomedin is involved in mammary growth regulation.

Although the type II is greater than the type I receptor population in mammary tissue during pregnancy, IGF-II is much less potent than IGF-I in inducing cell division. The population of IGF-I receptors during established lactation appears to be small. IGF-I is not as potent as insulin in inducing casein synthesis from mid-pregnant mammary tissues ; however, the effects of these two hormones are additive.

Exogenous treatment of lactating cows with BST results in increased plasma and milk IGF-I concentrations. However, the concentration of IGF-I in milk of BST-treated animals is well within the range of milk somatomedin concentrations detected throughout lactation. In fact, season and days open affect milk IGF-I concentration as much as or more than treatment of cattle with BST.

Insulin-like growth factor II binds to the type I and type II receptors. Neither insulin, nor IGF-I compete with IGF-II for the type II receptors at physiological concentrations of these molecules when recombinant molecules are used as the ligands. In contrast, IGF-II does compete for the type I somatomedin receptors. Type II receptors are the dominant form of somatomedin receptor during pregnancy and early lactation, and this is reflected in the milk concentration of this hormone. The concentration of IGF-II in mammary secretion is also elevated during pregnancy and then it falls with the onset of lactation : this may be associated with a smaller population of type II receptors during established lactation.

It should be stressed that exogenous administration of somatotropin to lactating cattle causes an increase in IGF-II concentrations in plasma but not in milk. To date, however, there are no specific studies on the biological responses of lactating mammary tissue to IGF-II.

To increase the milk yield, the practical utilization of IGF-I is limited, probably because of its scarcity and because, as a peptide circulating in high concentration, both the dose required and the mode of administration remain an open problem. In any case, it is useful to study the potential utilization of IGF-I in situations where the lipolytic and/or the anti-lipogenic effects of BST are not required, since these are thought to be mediated via direct effect of somatotropin on adipose tissue.

From a practical point of view, it seems possible to overcome the limitation of having to administer large amounts of IGF-I exogenously by enhancing the activity of endogenous IGF-I, rather than by increasing its concentration, as BST does. Circulating IGF-I is largely bound to specific proteins. The major binding protein species has a molecular weight of 150 kD, while some IGF-I is associated with a 50 kD protein. During BST treatment of lactating cows, the binding activity of the 150 kD protein species increases in parallel with the concentration of IGF-I. The association of IGF-I with these protein species restricts its movement out of the vascular space and attenuates its biological activity.

The possibility of using antibodies to prevent such interactions could serve to enhance IGF-I bioactivity. Furthermore, the truncated form of IGF-I fails to bind to the low-molecular-weight binding-protein and shows considerably enhanced biological activity *in vitro*. A binding-protein for growth hormone in human serum exists which may also be susceptible to manipulation in order to enhance IGF-I activity.

## ALTERNATIVE MANIPULATIONS BY ANTIBODIES

Monoclonal antibodies when complexed to somatotropin, rather than neutralizing its activity, actually enhance its potency. The mechanism of such an effect could be related to increased half-life, enhanced affinity for somatotropin receptors, or selective enhancement of the ability to bind to certain types of somatotropin receptors. It should, however, be stressed that recent studies also show enhanced activity of somatotropin when it is conjugated to albumin or indeed to itself.

An alternative approach is the possible use of antibodies to adipocytes, which are capable of destroying such cells *in vivo*, thus considerably reducing body fat deposition. Antibodies to adipocytes increase liveweight gain and food conversion efficiency, while long-term localized reduction of subcutaneous fat occurs.

Under the neuroendocrine control of the anterior pituitary, the pancreas is involved in the biosynthesis of several polypeptide hormones which regulate the metabolism of glucose and other major nutrients. This function is carried out by clusters of specialized cells, the *islets of Langerhans*, which contain several different types of related cells, each type forming a single kind of pancreatic hormone. The well-known hormones secreted by the islets are *glucagon*, made by the A cells ; *insulin*, made by the B cells ; *somatostatin*, made by the D cells, etc..

The immune system may be utilized to ablate pituitary somatotrophs selectively, leading to long term reduction in body weight gain. An alternative approach has been to target toxins specifically to pituitary cells, using the appropriate hypothalamic releasing factor conjugated to the toxin. Antibodies to somatostatin could conceivably be used in this way to destroy somatostatin secreting cells and, thereby, enhance endogenous somatotropin release.

The possible use of the anti-idiotypic network of the immune system to manufacture growth hormone "look-alikes" is a potential alternative to BST. Anti-idiotypes which mimic insulin action and β-adrenergic agents are available in the lab. Naturally-occurring autoantibodies with hormonal effects are also available in the lab, such as long-acting thyroid-stimulator (LATS), an immunoglobulin which binds to the TSH receptor.

Anti-idiotypic antibodies to somatotropin (which specifically displace labelled somatotropin from somatotropin-receptors in liver and adipose tissue, and which increase body-weight gain in hypophysectomized animals) are produced in the lab. Anti-idiotypes could be induced by immunization, using the idiotypic antibody, which would lead to increased growth hormone images.

Furthermore, it should be possible to produce in the lab monoclonal anti-idiotypes which mimic only a portion of the somatotropin molecule : this would permit the examination of the possibility that different parts of the somatotropin molecule possess different bioactivities.

## ALTERNATIVE MANIPULATIONS BY β-AGONISTS

The β-agonists (in particular cimaterol and ractopamine) show a marked ability to augment lean tissue whilst reducing fat deposition. These compounds :
 (a) exert direct effects on muscle protein synthesis and degradation ;
 (b) are orally active and can hence be incorporated into feed stuffs ;
 (c) are active in humans, making their handling more of a problem ;
 (d) may have deleterious effects on meat quality.

## ALTERNATIVE MANIPULATIONS BY SOMATOTROPIN FUSION GENES

The technology for introducing cloned genes into animals has been available only since 1980, and animals that have integrated a cloned gene into the genome are called "transgenic". The insertion of genes into animals has become extraordinarily useful for studies of gene function, developmental biology, physiology, pharmacology, and immunology.

The potential power that gene transfer offers is most dramatically demonstrated by the production of the well-known transgenic *supermouse*, which is the consequence of the secretion of a high concentration of rat growth hormone. It is possible to use the regulatory (promoter) sequence of a metallothionein gene to direct the transcription of the somatotropin structural gene and, thereby, avoid the usual hormonal feedback mechanism that regulates somatotropin.

The increase in somatotropin results in increased feed efficiency, lean muscle mass, increased milk production, along with reduced subcutaneous fat. Thus, interest in the production of transgenic animals with somatotropin fusion genes has been intense. Transgenic animals can be produced by retrovirus infection, or by introducing transformed embryonic stem cells into the blastocyst. To date microinjection of DNA into a pronucleus or nucleus has been used successfully in animals.

This transgenic approach inevitably results in concern for transgenic animal welfare and has not been without its early problems. Exposure to a very high concentration of somatotropin over prolonged periods changes not only the body composition but also body proportions. For example, liver size increases dramatically to a much greater extent that total body size. In transgenic mice, for example, all organs (with the exception of the brain) are increased relative to the increase in body size.

# 9

# Final remarks on BST activity

## GALACTOPOIETIC EFFECT AND NUTRIENT PARTITIONING

BST is capable of enhancing the milk yield to some extent. This effect is related to the modified nutrient utilization and mobilization in non-mammary tissues, with a sparing action of essential nutrients for milk synthesis.

Milk is produced both by direct intervention on the peripheral tissues and by changes in the responsiveness of the tissues to other metabolic hormones, namely insulin and catecholamines. In the mammary gland, BST seems to stimulate milk synthesis via insulin-like growth factor I (IGF-I).

During the initial period after BST treatment, cows are likely to move into negative energy balance and so a mobilization of body lipids is required. In lactating animals this fatty acid mobilization both supplies a source of precursor for milk synthesis and provides a means of decreasing glucose oxidation by peripheral tissues.

Subsequently, the BST action results in an increased food intake and so the need to mobilize fat reserves is removed because milk yield can be fuelled by extra-dietary nutrients.

The percentage increase in lactation yield by BST treatment can vary from less than 5 % to up to 25 %, depending on the genetic potential of the cows, and their feeding and management conditions. BST responses among adult cows are highly variable and there is no constant difference in the BST responses between heifers and multiparous cows.

Percentage of milk response to BST can decrease when milk yield increases. The enhancement of feed efficiency results :

(a) from the dilution of maintenance requirement in the total requirement, due to increased milk production, and

(b) from less use of nutrients for body tissue deposition relative to milk secretion.

## INFLUENCE ON COW HEALTH, WELFARE AND REPRODUCTION

In lactating dairy cows, BST does not seem to significantly change the incidence of clinical diseases (including mastitis), bone development, Ca and P metabolism, and so on. There is no sign of either subclinical or clinical ketosis, milk fever or burn-out, although there are some significant alterations in milk somatic cell count and blood chemistry (shift in the balance between the total number of red cells and the total volume of plasma).

High dose rates of BST in early lactation tend to diminish reproductive efficiency to that sometimes seen in naturally high-yielding cows, probably because of the negative energy balance caused by the higher milk yield. Calf health, growth and subsequent development seem to be unaffected by BST treatment.

If the plane of nutrition is inadequate or management conditions are bad, the cow's response to BST will be less because of the negative energy balance and an increase in disease-related incidences could occur.

Adequate management must be available to provide appropriate facilities for BST injection. Furthermore, adequate veterinary supervision must be constantly available to avoid BST abuse or misuse, namely :

— the administration of BST to exploit overstretched cull cows by extending lactation ;

— the administration of BST without adequate nutritional levels ;

— the administration of BST to cows whose milk yield is depressed by undiagnosed disease or metabolic disorder.

The subcutaneous or intramuscular injection of a small quantity of BST material every two or four weeks is not more painful than any other injection. Certain sites of the cows may be less sensitive than others and, therefore, clear guidelines for injection sites must be drawn up.

The transient tissue swelling from injections is resolved in 2-4 days with the intramuscular route, and in 2-4 weeks with the subcutaneous route. Compulsory adequate hygienic standards prevent abscesses or complications.

Genetically selected high-yielding cows suffer energy deficit during the post-partum, with subsequent adverse effects on fertility, incidence of ketosis, etc.. Because milk yield and persistency are negatively correlated, BST treatment would make it possible to put less emphasis on initial milk production. That is, if in genetic selection less emphasis needs to be put on milk yield, more emphasis can be placed on resistance to diseases, better temperament, lower incidence of lameness, and so on.

## INFLUENCE ON MILK CHARACTERISTICS AND PROCESSING SUITABILITY

The milk protein content increases slightly during the fourth week after BST injection, while during the first week the fat content increases together with the milk yield, and both decrease afterwards. Generally the casein content and the casein composition are not markedly influenced by BST treatment. To date, however, there are few data on the effect on the individual whey proteins which play a role in milk processing.

After BST treatment the milk concentrations of Ca, P, Na, Fe, Cu, Zn, Mg and Mn do not seem to be affected. Similar results are observed for milk enzyme activities, such as aminopeptidase, endopeptidase, phosphatase, lipase and total protease. Unfortunately, to date there are no data on the individual protease types. The limited available data seem to indicate that BST treatment causes no changes in the milk vitamin concentrations.

Higher values of both pyruvate content and somatic cell count can occur in milk from BST-treated cows. This topic is very important because in the European Community these parameters are used as a part of controlling milk quality.

## ALTERATIONS IN BLOOD AND MILK SOMATOTROPIN CONCENTRATIONS

In cows absolute *blood* concentrations of somatotropin increase significantly during treatment with prolonged-release BST. The plasma levels of somatotropin are within the range of growing animals, but the higher concentrations last longer than the relatively brief episodes of endogenous somatotropin release.

Bovine somatotropin is secreted into *milk* of untreated cows only at minimal detectable concentrations below 0.5 ng/ml skim milk. Treatment of cows (performed according to the suggested doses and rate of BST administration) does not seem to increase the somatotropin concentrations in milk.

## ALTERATIONS IN BLOOD AND MILK IGF-I CONCENTRATIONS

Absolute *blood* values for IGF-I increase 2-5 fold after BST injection, but plasma levels seem to remain within the range observed in growing animals. After the conclusion of treatment, the elevated IGF-I levels decrease to the control values within a relatively short time.

During BST treatment at recommended doses, a small increase of IGF-I in *milk* is present, particularly after the second and third injections. Although the IGF-I levels in milk of BST-treated cows are within the range of values detected in milk from untreated cows, the similarity between bovine IGF-I and the human insulin-like growth factors could cause concern.

From a speculative point of view, however, it does not seem that to date this would represent a human health hazard because :

(a) the milk IGF-I concentrations are small and within the range values detected in the milk of untreated cows and human breast milk ;

(b) such peptides would be broken down and inactivated by milk pasteurization, or sterilization, or processing ;

(c) such peptides would be readily digested in the human intestinal tract ;

(d) effects of IGF-I administered to humans have only been noted when large doses have been given parenterally.

## RESIDUAL CONCENTRATIONS IN MEAT

At this time there is no known technology that can assay BST levels in muscle or other body tissues (other than blood and milk). This is because it is at present impossible to extract and separate the 191 amino acid chain, intact with its structural relationship unaltered, from the other tissue proteins.

The concept of labelling or tagging the BST molecule has been discussed many times but unfortunately it is not a practical proposition. This is because both naturally-occurring pituitary and recombinant BST have very short half-lives in the body and, therefore, the BST molecule is rapidly broken down in the body and its amino acids are utilized, for example, to make other body proteins.

To make up for the absence of empirical data, by using an oversimplified arithmetical computation it is possible to show that, after BST treatment at recommended doses, the theoretical concentration of BST in *meat* would be markedly low. Thus, even assuming that BST is not broken down in the human gut, to reach an active pharmacological dosage a man would have to eat enormous amounts of meat each day.

BST would be readily digested in the human intestinal tract and, even if some small amounts of intact bovine somatotropin could be absorbed into the human body, it does not seem likely that it could produce significant effects in humans for two reasons. In the first instance somatotropins seem to be species-specific and bovine somatotropin, injected parenterally, is without effect in humans. Additionally, minute amounts of growth hormone, even if absorbed intact after oral ingestion, would be subject to normal metabolic breakdown and seem unlikely to produce significant effects in humans.

As to the possibility that fragments derived from bovine somatotropin may have an effect in humans, it seems that this is most unlikely as well. For example, one biologically active peptide (i.e. bGH 96-133) is derived from bovine pituitary somatotropin and does have activity in humans. However, if it appears in the gut, this peptide is more rapidly degraded than intact growth hormone itself. In addition, even when this material is administered parenterally, it is necessary to give very large amounts to demonstrate an effect in hypopituitary humans.

These issues are essentially the same for recombinant bovine somatotropin. Its amino acid composition and sequence are essentially homologous to pituitary somatotropin and studies with recombinant bovine somatotropin have not demonstrated any significant differences in biological responses. Thus, from a biological point of view, it should not be considered the recombinant somatotropin is different from pituitary somatotropin. The same considerations apply to peptides derived therefrom, because chemically the same peptides are derived from pituitary or recombinant bovine somatotropin.

## INFLUENCE ON MEAT CHARACTERISTICS

To date, very little evidence is available on the effects of BST administration on meat quality and none of the studies in cattle could be described as large-scale and long-term trials.

One trial, with the largest number of animals per treatment, shows a significant effect on taste panel scores for tenderness and overall acceptability and small differences have been found in the adverse direction, namely, shear force, collagen solubility, collagen content, juice extraction and weight-loss during cooking. Further data will be necessary to resolve these issues.

*Biomedical problems on bovine somatotropin use in milk production.*
Ed. G. Benzi. John Libbey Eurotext. Paris © 1990, pp. 77-104.

# 10

# Essential bibliography

Abdel-Meguid S.S., Shieh H.-S., Smith W.W., Dayringer H.E., Violand B.N. and Bentle, L.A. (1987) **Three-dimensional structure of a genetically engineered variant of porcine growth hormone.** Proc. Natl. Acad. USA 84, 6434-6437.

Adam J.J., Fabry J., Ettaib A. and Deroanne, C. (1985) **Effect of exogenous bovine growth hormone upon Belgian white blue heifer's meat.** Rec. Méd. Vét. 161, 655-683.

Aguilar A.A., Jordan D.C., Olson J.D., Bailey C. and Hartnell, G.E. (1988) **A short-study evaluating the galactopoietic effects of the administration of Sometribove (recombinant methionyl bovine somatotropin) in high producing dairy cows milked three times per day.** J. Dairy Sci. 71 (Suppl. 1), 208.

Akers R.M. (1985) **Lactogenic hormones : binding sites, mammary growth, secretory cell differentiation and milk biosynthesis in ruminants.** J. Dairy Sci. 68, 501-509.

Akers R.M. and Keys J.E. (1984) **Characterization of lactogenic hormone binding to membranes from ovine and bovine mammary gland and liver.** J. Dairy Sci. 67, 2224-2235.

Allen P. and Enright W.J. (1989) **Effects of administration of somatotropin on meat quality in ruminants : a review.** In "Use of Somatotropin in Livestock Production" (Eds. K. Sejrsen M. Vestergaard and A. Neimann-Srensen), Elsevier Applied Science, London and New York, pp. 201-209.

Anderson M.J., Lamb R.C., Arambel M.J., Boman R.L., Hard D.L. and Kung L. (1988) **Evaluation of a prolonged release system of sometribove, USAN (recombinant methionyl bovine somatotropin) on feed intake, body weight, efficiency and energy balance of lactating cows.** J. Dairy Sci. 71 (Suppl. 1), 208.

Annexstad R.J., Otterby D.E., Linn J.G., Hansen W.P., Soderholm C.G. and Eggert R.G. (1987) **Responses of cows to daily injections of recombinant bovine somatotropin (BST) during a second consecutive lactation.** J. Dairy Sci. 70 (Suppl. 1), 176.

Anonymous (1984) **Diet and cardiovascular disease. Department of Health and Social Security. Report on Health and Social Aspects,** No. 28. H.M.S.O., London.

Asimov G.J. and Krouze N.K. (1937) **The lactogenic preparations from the anterior pituitary and the increase of milk yield in cows.** J. Dairy Sci. 20, 289-306.

Aston R., Holder A.T., Preece M.A. and Ivanyi J. (1986) **Potentiation of the somatogenic and lactogenic activity of human growth hormone with monoclonal antibodies.** J. Endocr. 110, 381-388.

Auberger B., Lenoir J. and Remeuf, F. (1988) **L'incidence du traitement de vaches laitières par la somatotropine bovine sur la composition et les aptitudes technologiques du lait.** Techn. Lait. Mark. 1030, Technologie 1-3.

Baer R.J., Treszen K.M., Schingoethe D.J., Caspar D.P., Shaver R.D. and Cleale, R.M. (1988) **Composition and flavour of milk produced by cows injected with recombinant bovine somatotropin.** J. Dairy Sci. 71 (Suppl. 1), 115.

Baile C.A. and Krestel-Rickert, D.H. (1985) **Somatotropin. Feed Management** 36, 26-32.

Baird L.S., Hemken R.W., Harmon R.J. and Eggert R.G. (1986) **Response of lactating dairy cows to recombinant bovine growth hormone (rbGH).** J. Dairy Sci. 69 (Suppl. 1), 118.

Ballard F.J., Ross M., Upton F.M. and Francis G.L. (1988) **Specific binding of insulin-like growth factors I and II to the type I and II receptors, respectively.** Biochem. J. 249, 721-726.

Barbano D.M. Lynch J.M., Bauman D.E. and Hartnell G.F. (1988) **Influence of sometribove (recombinant methionyl bovine somatotropin) on general milk composition.** J. Dairy Sci. 71 (Suppl. 1), 101.

Barenton B., Guyda H.J., Goodyer C.G., Polychronakos C. and Posner B.I. (1987) **Specificity of insulin-like growth factor binding to type-II IGF receptors in rabbit mammary gland and hypophysectomized rat liver.** Biochem. Biophys. Res. Comm. 149, 555-561.

Barnard R., Bundesen P.G., Rylatt D.R. and Waters M.J. (1985) **Evidence from the use of monoclonal antibody probes for structural heterogeneity of the hormone receptor.** Biochem. J. 231, 459-468.

Bauman D.E. (1984) **Regulation of nutrient partitioning.** In "Herbivore Nutrition in the Subtropics and Tropics" (Eds. F.M.C. Gilchrist and R.I. Machie), The Science Press, Craighall, South Africa, pp. 505-524.

Bauman D.E. (1987) **Bovine somatotropin. The Cornell experience.** In "National Invitational Workshop on Bovine Somatotropin". St. Louis, USA, Sept. 21-23.

Bauman D.E. and Currie W.B. (1980) **Partitioning of nutrients during pregnancy and lactation : a review of mechanism involving homeostasis and homeorhesis.** J. Dairy Sci. 63, 1514-1529.

Bauman D.E. and McCutcheon S.N. (1986) **The effects of growth hormone and prolactin on metabolism.** In "Control of Digestion and Metabolism in Ruminants" (Eds. L.P. Milligan W.L. Grovum and A. Dobson). Proceedings of the Sixth International Symposium on Ruminant Physiology. Prentice-Hall, New Jersey, pp. 436-455.

Bauman D.E., DeGeeter M.J., Peel C.J., Lanza G.M., Gorewit R.C. and Hammond R.W. (1982) **Effects of recombinantly derived bovine growth hormone (bGH) on lactational performance of high yielding dairy cows.** J. Dairy Sci. 65 (Suppl. 1), 121.

Bauman D.E., Eppard P.J. and McCutcheon S.N. (1984) **Effects of exogenous somatotropin in lactating dairy cows.** In "New Trends in Animal Nutrition and Physiology". Monsanto International Symposium. Louvain-la-Neuve, Belgium.

Bauman D.E., Eppard P.J., DeGeeter M.J. and Lanza G.M. (1985) **Responses of high producing dairy cows to long-term treatment with pituitary somatotropin and recombinant somatotropin.** J. Dairy Sci. 68, 1352-1362.

Bauman D.E., Hard D.L., Crooker B.A., Erb H.N. and Sandles L.D. (1988a) **Lactational performance of dairy cows treated with a prolonged-release formulation of methionyl bovine somatotropin (sometribove).** J. Dairy Sci. 71 (Suppl. 1), 205.

Bauman D.E., Peel C.J., Steinhour W.D., Reynolds P.J., Tyrrell H.F., Brown A.C.G. and Haaland G.L. (1988b) **Effect of bovine somatotropin on metabolism of lactating dairy cows : influence on rates of irreversible loss and oxidation of glucose and nonesterified fatty acids. J. Nutr. 118, 1031-1040.

Baumann G., Stolar M.W., Amburn K., Barsano C.P. and de Vries B.C. (1986) **A specific growth hormone-binding protein in human plasma. Initial characterization.** J. Clin. Endocr. Metab. 62, 134-141.

Baumrucker C.R. (1985) **Growth hormone does not directly affect bovine mammary tissue growth nor lactating acini milk production in culture.** J. Dairy Sci. 69 (Suppl. 1), 106.

Baxter R.C., Zaltsman Z. and Turtle J.R. (1984a) **Rat growth hormone (GH) but not prolactin (PRL) induces both GH and PRL receptors in female rat liver.** Endocr. 114, 1893-1901.

Baxter R.C., Zaltsman Z. and Turtle J.R. (1984b) **Immunoreactive somatomedin-C/insulin-like growth factor I and its binding protein in human milk.** J. Clin. Endocr. Metab. 58, 955-959.

Bechtel P.J., Easter R.A., Novakofski J., McKeith F.K., McLaren D.G., Jones R.W. and Ingle D.L. (1988) **Effect of porcine somatotropin on limit fed swine.** J. Anim. Sci. 66 (Suppl. 1), 282.

Beck J.C., McGarry E.E., Dyrenfurth I. and Venning E.H. (1957) **Metabolic effects of human and monkey growth hormone in man.** Science 125, 384-385.

Beckers J.F. (1987) **Coexistence of lactogenic and somatogenic receptors in bovine mammary gland.** Proceedings of the 1st European Congress of Endocrinology, June 21-25, p. 138.

Beerman D.H., Hogue D.E., Fishell V.K., Dickson H.W., Aronica S., Dwyer D. and Schricher, B.R. (1988) **Effects of exogenous ovine growth hormone (oGH) and human GH releasing factor (hGRF) on plasma oGH concentration and composition of gain in lambs.** J. Anim. Sci. 66 (Suppl. 1), 282-283.

Bell A.W. (1980) **Lipid metabolism in liver and selected tissues and in the whole body of ruminant animals.** Prog. Lipid Res. 18, 117-164.

Bergen W.G., Johnson S.E., Skjaerlund D.M., Merkel R.A. and Anderson D.B. (1987) **The effect of ractopamine on skeletal muscle metabolism in pigs.** Fed. Proc. 46, 1021.

Bergenstal D.M. and Lipsett M.B. (1960) **Metabolic effects of human growth hormone and growth hormone of other species in man.** J. Clin. Endocr. Metab. 20, 1427-1436.

Bines J.A. and Hart I.C. (1982) **Metabolic limits to milk production, especially roles of growth hormone and insulin.** J. Dairy Sci. 65, 1375-1389.

Bines J.A., Hart I.C. and Morant S.V. (1980) **Endocrine control of energy metabolism in the cow : the effect on milk yield and levels of some blood constituents of injecting hormone and growth hormone fragments.** Br. J. Nutr. 43, 179-188.

Birmingham B.K., White T.C., Lanza G.M., Miller M.A., Torkelson A.R. and Hale M.D. (1988) **Pharmacokinetics of sometribove, USAN (recombinant methionyl bovine somatotropin) and a naturally occurring somatotropin variant in lactating dairy cows.** J. Dairy Sci. 71 (Suppl. 1), 194.

Bitman J., Wood D.L., Tyrrell H.F., Bauman D.E., Peel C.J., Brown A.C.G. and Reynolds P.J. (1984) **Blood and milk lipid responses induced by growth hormone administration in lactating cows.** J. Dairy Sci. 67, 2873-2880.

Bonczek, R.R., Young C.W., Wheaton J.E. and Miller K.P. (1988) **Response of somatotropin, insulin, prolactin and thyroxine to selection for milk yield in Holsteins.** J. Dairy Sci. 71, 2470-2479.

Booth J.M. (1988) **Progress in controlling mastitis in England and Wales.** Vet. Rec. 122, 299-302.

Bourchier C.P., Garnswoth P.C., Hutchinson J.M. and Benson T.A. (1987) **The relationship between milk yield, body condition and reproductive performance in high yielding dairy cows.** Anim. Prod. 41, 460.

Boyd R.D. and Bauman D.E. (1988) **Mechanism of action for somatotropin in growth.** In "Current Concepts of Animal Growth Regulation" (Eds. D.R. Campion, G.J. Hausman and R.J. Martin), Plenum, New York.

Boyd R.D., Bauman D.E., Beermann D.H., De Neergaard A.F., Souza L. and Butler W.R. (1986) **Titration of the porcine growth hormone dose which maximises growth performance and lean deposition in swine.** J. Anim. Sci. 63 (Suppl. 1), 218.

Boyd R.D., Beerman D.H., Roneker K.R., Bartley T.D. and Fagin K.D. (1988) **Biological activity of a recombinant variant (21 kd) of porcine somatotropin in growing swine.** J. Anim. Sci. 66 (Suppl. 1), 256.

Braithwaite G.D. (1975) **The effect of growth hormone on calcium metabolism in the sheep.** Br. J. Nutr. 33, 309-314.

Breier B.H., Bass J.J., Butler J.H. and Gluckman P. (1986) **The somatotrophic axis in young steers : influence of nutritional status on pulsatile release of growth hormone and circulating concentrations of insulin-like growth factor I.** J. Endocr. 111, 209-215.

Breier B.H. Gluckman P.D. and Bass J.J. (1988a) **Influence of nutritional status and oestradiol-17b on plasma growth hormone, insulin-like growth factors -I and -II and the response to exogenous growth hormone in young steers.** J. Endocr. 118, 243-250.

Breier B.H., Gluckman P.D. and Bass J.J. (1988b) **The somatotrophic axis in young steers : influence of nutritional status and oestradiol-17b on hepatic high- and low-affinity somatotrophic binding sites.** J. Endocr. 116, 169-177.

Brems D.N., Plaisted S.M., Havel H.A., Kauffman E.W., Stodola J.D., Eaton L.C. and White R.D. (1985) **Equilibrium denaturation of pituitary- and recombinant-derived bovine growth hormone.** Biochemistry 24, 7662-7668.

Brenner K.-V., Novakofski J., Bechtel P.J. and Easter R.A. (1989) **Metabolic and endocrine challenge of somatotropin treated pigs.** In "Biotechnology in Growth Regulation" (Eds. R.B. Heap, C.G. Prosser and G.E. Lamming), Butterworth, London.

Brown D.L., Taylor S.J., De Peters E.J. and Baldwin R.L. (1988) **Influence of sometribove (a methionyl bovine somatotropin) on the body composition of lactating Holstein cattle.** J. Dairy Sci. 71 (Suppl. 1), 125.

Brumby P.J. (1959) **The influence of growth hormone on growth in young cattle.** N. Z. J. Agric. Res. 2, 683-686.

Brumby P.J. and Hancock J. (1955) **The galactopoietic role of growth hormone in dairy cattle.** N. Z. J. Tech. 36, 417-436.

Bryan K.A., Carbaugh D.E., Clark A.M., Hagen D.R. and Hammond J.M. (1987) **Effect of porcine growth hormone (pGH) on growth and carcass composition of gilts.** J. Anim. Sci. 65 (Suppl. 1), 244.

Bryan K.A., Clark A.M. and Hagen D.R. (1988) **Effect of treatment with and withdrawal of porcine growth hormone (pGH) on growth and reproductive performance of gilts.** J. Anim. Sci. 66 (Suppl. 1), 401.

Burnside E.B. and Meyer K. (1987) **Effect of different strategies of administration of bovine somatotropin on within herd variance and accuracy of sire rankings for milk production.** J. Dairy Sci. 70 (Suppl. 1), 127.

Burridge M.J. (1979) **Review of epidemiologic studies investigating possible association between Bovine Leukosis and human disease.** In "Proceedings of Bovine Leukosis Symposium", USDA Animal and Plant Health Inspection Service, Veterinary Services and Science and Education Administration, pp. 135-142.

Burton J.H., McBride B.W., Bateman K., MacLeod G.K. and Eggert R.G. (1987) **Recombinant bovine somatotropin : effects on production and reproduction in lactating cows.** J. Dairy Sci. 70 (Suppl. 1), 175.

Caldwell G.G. (1979) **Bovine Leukemia Virus, public health serologic studies.** In "Proceedings of Bovine Leukosis Symposium", USDA Animal and Plant Health Inspection Service, Veterinary Services and Science and Education Administration, pp. 143-159.

Campbell P.G. and Baumrucker C.R. (1988) **Secretion of immunoreactive insulin-like growth factor I and its binding protein from the bovine mammary gland in vitro.** Proc. Endocr. Soc. 70th Annual Meeting, Abst. 510.

Campbell R.G. and Travener M.R. (1988) **Genotype and sex effects on the responsiveness of growing pigs to exogenous porcine growth hormone (pGH) administration.** J. Anim. Sci. 66 (Suppl. 1), 257.

Campbell R.G., Caperna T.J., Steele N.C. and Mitchell A.D. (1987) **Effect of porcine pituitary growth hormone (pGH) administration and energy intake on growth performance of pigs from 25 to 55 kg body weight.** J. Anim. Sci. 65 (Suppl. 1), 244.

Campbell R.G., Steele N.C., Caperna T.J., McMurtry J.P., Solomon M.B. and Michell A.D. (1988) **Interrelationships between energy intake and endogenous porcine growth hormone administration on the performance, body composition and protein and energy metabolism of growing pigs weighing 25 to 55 kilograms liveweight.** J. Anim. Sci. 66, 1643-1655.

Campbell R.M., Su C.-M., Maines S.L., Stricker P.R., Jensen L.R., Heimer E.P., Felix A.M. and Mowles T.F. (1988) **Biological activities of novel cyclic growth hormone-releasing factor (GRF) analogs.** J. Anim. Sci. 66 (Suppl. 1), 291.

Caperna T.J., Campbell R.G. and Steele N.G. (1987) **Influence of porcine growth hormone (pGH) administration and energy intake on mineral status of growing swine.** J. Anim. Sci. 65 (Suppl. 1), 254.

Car M., Znidar A. and Filipan T. (1967) **An effect of the treatment of young steers with STH (growth hormone) upon nitrogen retention in intensive feeding.** Veterinarski arhiv. 5-6, 173-184.

Castro-Magaña M., Angulo M., Fuentes B., Castelar M.E., Cañas A. and Espinoza B. (1986) **Effect of prolonged clonidine administration on growth hormone concentrations and rate of linear growth in children with constitutional growth delay.** J. Pediatr. 109, 784-787.

Chaiyabutr N., Faulkner A. and Peaker M. (1983) **Effects of exogenous glucose on glucose metabolism in the lactating goat in vivo.** Br. J. Nutr. 49, 159-165.

Chalupa W. and Galligan D.T. (1988) **Nutritional implications of somatotropin for lactating cows.** J. Dairy Sci. 71 (Suppl. 1), 123.

Chalupa W., Hausman B., Kronfeld D.S., Kensinger R.S., McCarthy R.D. and Rock D.W. (1984) **Responses of lactating cows to exogenous growth hormone and dietary sodium bicarbonate. I. Production.** J. Dairy Sci. 67 (Suppl. 1), 107.

Chalupa W., Vecchiarelli B., Schneider P. and Eggert R.G. (1986) **Long-term responses of lactating cows to daily injection of recombinant somatotropin.** J. Dairy Sci. 69 (Suppl. 1), 151.

Chalupa W., Baird L., Soderholm C., Palmquist D.L., Hemken R., Otterby D., Annexstad R., Vecchiarelli B., Harman R., Sinha A., Linn J., Hansen W., Ehle F., Schneider P. and Eggert R.G. (1987a) **Response of dairy cows to somatotropin.** J. Dairy Sci. 70 (Suppl. 1), 176.

Chalupa W., Marsh W.E. and Galligan D.T. (1987b) **Bovine somatotropin : lactational responses and impacts on feeding programs.** Proc. Maryland Nutr. Conf. Feed Manuf., pp. 48-57.

Chalupa W., Kutches A., Swager D., Lehenbauer T., Vecchiarelli B., Shaver R. and Robb E. (1988) **Responses of cows in a commercial dairy to somatotropin.** J. Dairy Sci. 71 (Suppl. 1), 210.

Chew B.P., Eisenman J.R. and Tanaka T.S. (1984) **Arginine infusion stimulates prolactin, growth hormone, insulin, and subsequent lactation in pregnant dairy cows.** J. Dairy Sci. 67, 2507-2518.

Chilliard Y. (1987) **Revue bibliographique. Variations quantitatives et métabolisme des lipides dans les tissus adipeux et le foie au cours du cycle gestation-lactation. 2ème partie : chez la brebis et la vache.** Reprod. Nutr. Dévelop. 27, 327-398.

Chilliard Y. (1988) **Rôles et mécanismes d'action de la somatotropine (hormone de croissance) chez le ruminant en lactation.** Reprod. Nutr. Dévelop. 28, 39-59.

Chilliard Y. (1989) **Long-term effects of recombinant bovine somatotropin (rBST) on dairy cow performances : a review.** In "Use of Somatotropin in Livestock Production" (Eds. K. Sejrsen, M. Vestergaard and A. Neimann-Srensen), Elsevier Applied Science, London and New York, pp. 61-87.

Chilliard Y. and Robelin J. (1983) **Mobilization of body proteins by early lactating cows measured by slaughter and D20 dilution techiques.** IVth Int. Symp. Protein metabolism and nutrition (Clermont-Ferrand), EAAP - Publ. N 31, V I. II, 195-198 (INRA Publ.).

Chung C.S., Etherton T.D. and Wiggins J.P. (1985) **Stimulation of swine growth by porcine growth hormone.** J. Anim. Sci. 60, 118-130.

Chung Y.S., Prior H.C., Duffy P.F., Rogers R.J. and Mackenzie A.R. (1986) **The effect of pasteurisation on Bovine Leukosis Virus-infected milk.** Aust. Vet. J. 63, 379-380.

Cohen S.N., Chang A.C.Y., Boyer H.W. and Helling R.B. (1973) **Construction of biologically functional bacterial plasmids in vitro.** Proc. Natl. Acad. Sci. USA 70, 3240.

Cohick W.S., Slepetis R., Plaut K. and Bauman D.E. (1987) **Effect of exogenous somatotropin on serum somatomedin-C (SmC) and hepatic metabolism of lactating cows.** J. Anim. Sci. 65 (Suppl. 1), 248.

Cole W.J., Eppard P.J., Lanza G.M., Hintz R.L., Madsen K.S., Franson S.E., White T.C., Ribelin W.E., Hammond B.G., Bussen S.C., Leak R.K. and Metzger L.E. (1988) **Response of lactation dairy cows to multiple injections of sometribove, USAN (recombinant methionyl bovine somatotropin) in a prolonged release system. Part II. Health and reproduction.** J. Dairy Sci. 71 (Suppl. 1), 184.

Collier R.J. and Johnson H.D. (1988) **Bovine somatotropin. Mechanism of action and effects under differing environments.** Monsanto technical Symposium, Fresno, California, USA, pp. 11-19.

Collier R.J., Li W.R., Johnson H.D., Becker B.A., Buonomo F.C. and Spencer K.J. (1988) **Effect of sometribove (methionyl bovine somatotropin, BST) plasma insulin-like growth factor I (IGF-I) and II (IGF-II) in cattle exposed to heat and stress.** J. Dairy Sci. 71 (Suppl. 1), 228.

Corps A.N., Brown K.D., Rees L.H., Carr J. and Prosser C.G. (1988) **The insulin-like growth factor I content in human milk increases between early and full lactation.** J. Clin. Endocr. Metab. 67, 25-29.

Crooker B.A., Bauman D.E., Cohick W.S. and Harkins W. (1988) **Effect of dose of exogenous bovine somatotropin on nutrient utilization by growing dairy heifers.** J. Anim. Sci. 66 (Suppl. 1), 299.

Croom Jr., W.J., Leonard E.S., Baker P.K., Kraft L.A. and Ricks, C.A. (1984) **The effects of synthetic growth hormone releasing hexapeptide. BI 679 on serum growth hormone levels and production in lactating dairy cattle.** J. Dairy Sci. 67 (Suppl. 1), 109.

## Essential bibliography

Darnell J., Lodish H. and Baltimore D. (1986) "Molecular Cell Biology". W.H. Freeman New York.

Daughaday W.H. (1981) **Growth hormone and the somatomedins.** In "Endocrine Control of Growth" (Ed. W.H. Daughaday), Elsevier, New York, pp. 1-24.

Daughaday W.H. (1985) **Prolactin and growth hormone in health and disease.** In "Contemporary Endocrinology" (Ed. S.H. Ingbar), Plenum Publ. Co., New York, vol. 2, pp. 27-86.

Davidson M.B. (1987) **Effect of growth hormone on carbohydrate and lipid metabolism.** Endocr. Rev. 8, 115-131.

Davis S.L., Garrigus U.S. and Hinds F.C. (1970) **Metabolic effects of growth hormone and diethylstilbestrol in lambs. II. Effects of daily ovine growth hormone injections on plasma metabolites and nitrogen-retention in fed lambs.** J. Anim. Sci. 30, 236-240.

Davis S.R., Gluckman P.D., Hart I.C. and Henderson H.V. (1987) **Effects of injecting growth hormone or thyroxine on milk production and blood plasma concentrations of insulin-like growth factors I and II in dairy cows.** J. Endocr. 114, 17-24.

Davis S.R., Collier R.J., McNamara J.P., Head H.H. and Sussman W. (1988a) **Effects of thyroxine and growth hormone treatment of dairy cows on milk yield.** J. Anim. Sci. 66, 70-79.

Davis S.R., Collier R.J., McNamara J.P., Head H.H., Croon W.J. and Wilcox C.J. (1988b) **Effects of thyroxine and growth hormone treatment of dairy cows on mammary uptake of glucose, oxygen and other milk fat precursors.** J. Anim. Sci. 66, 80-89.

Denton R.M., Hughes W.A., Bridges B.J., Brownsey R.W., McCormack J.G. and Stansbie D. (1978) **Regulation of mammalian pyruvate dehydrogenase by hormones.** Horm. Cell. Regul. 2, 191-208.

Desnouveaux R., Montigny H., Le Treut J.-H., Schockmel L. and Biju-Duval, B. (1988) **Vérification de l'aptitude fromagère du lait de vaches traitées à la somatotropine bovine méthionylée. Fabrication expérimentale de fromages à pates molles de type camembert.** Techn. Lait. Mark. 1030, Technologie 4-7.

Dhiman T.R., Kleinmans J., Radloff H.D., Tessmann N.J. and Satter L.D. (1988) **Effect of recombinant bovine somatotropin on feed intake, dry matter digestibility and blood constituents in lactating dairy cows.** J. Dairy Sci. 71 (Suppl. 1), 121.

Dickerson R.E. and Gies I. (1976) "The Structure and Function of Proteins". Harper & Row, New York.

Di Girolamo M., Edén S., Enberg G., Isaksson O., Lönron P, Hall K. and Smith U. (1986) **Specific binding of human growth hormone but not insulin-like growth factors by human adipocytes.** FEBS Lett. 205, 15-19.

Ducker M.J., Haggett R.A., Fisher W.J., Morant S.V. and Bloomfield G.A. (1985) **Nutrition and reproductive performance of dairy cattle.** Anim. Prod. 41, 1-12.

Eadara J., Dalrymple R.H., De Lay R.L., Ricks C.A. and Romsos D.R. (1987) **Cimaterol, a novel b-agonist, selectively stimulates white adipose tissue lipolysis and skeletal muscle lipoprotein lipase activity in rats.** Fed. Proc. 46, 1020.

Early R.J., McBride B.W., Ball R.O. and Rock D.W. (1988) **Growth, feed efficiency and carcass characteristics of beef steers treated with daily injections of recombinantly-derived bovine somatotropin.** J. Anim. Sci. 66 (Suppl. 1), 283.

Eastham E.J., Lichauco T., Grady M.I. and Walker W.A. (1978) **Antigenicity of infant formulas : role of immature intestine on protein permeability.** J. Pediatr. 93, 561-564.

Eisemann J.H., Hammond A.C., Bauman D.E., Reynolds P.J., McCutcheon S.N., Tyrrell H.F. and Haaland G.L. (1986a) **Effect of bovine growth hormone administration** on metabolism of growing hereford heifers : protein and lipid metabolism and plasma concentrations of metabolites and hormones. J. Nutr. 116, 2504-2515.

Eisemann J.H., Hammond A.C., Rumsey T.S. and Bauman D.E. (1986b) **Tissue protein synthesis rates in beef steers injected with placebo or bovine growth hormone.** J. Anim. Sci. 63 (Suppl. 1), 217.

Eisemann J.H., Tyrrell H.F., Hammond A.C., Reynolds P.J., Bauman D.E., Haaland G.L., McMurtry J.P. and Varga G.A. (1986c) **Effect of bovine growth hormone administration on metabolism of growing hereford heifers : dietary digestibility, energy and nitrogen balance.** J. Nutr. 116, 157-163.

Eisenbeisz W.A., Casper D.P., Schingoethe D.J. and Ludens F.C. (1988a) **Lactational evaluation of recombinant bovine somatotropin with corn and barley diets : response to diets.** J. Dairy Sci. 71 (Suppl. 1), 122.

Eisenbeisz W.A., Casper D.P., Schingoethe D.J., Ludens F.C. and Shaver R.D. (1988b) **Lactational evaluation of recombinant bovine somatotropin with corn and barley diets : response to somatotropin.** J. Dairy Sci. 71 (Suppl. 1), 123.

Elcock C., Buttle H.L., Coles H.J., Hathorn D.J., Simmonds A.D. and Pell J.M. (1988) **The effect of growth hormone on nutrient partitioning in lambs.** J. Endocr. 117 (Suppl. 1), 59.

Elvinger F., Head H.H., Wilcox C.J. and Natzke R.P. (1987) **Effects of administration of bovine somatotropin on lactation milk yield and composition.** J. Dairy Sci. 70 (Suppl.), 121.

Elvinger F., Head H.H., Wilcox C.J., Natzke R.P. and Eggert R.G. (1988) **Effects of administration of bovine somatotropin on milk yield and composition.** J. Dairy Sci. 71, 1515-1525.

Enright W.J. (1989) **Effects of administration of somatotropin on growth, feed efficiency and carcass composition of ruminants : a review.** In "Use of Somatotropin in Livestock Production" (Eds. K. Sejrsen, M. Vestergaard and A. Neimann-Srensen), Elsevier Applied Science, London and New York, pp. 132-156.

Enright W.J., Chapin L.T., Moseley W.M. and Tucker H.A. (1988) **Effects of infusions of various doses of bovine growth hormone-releasing factor on growth hormone and lactation in Holstein cows.** J. Dairy Sci. 71, 99-108.

Eppard P.J., Bauman D.E. and McCutcheon S.N. (1985a) **Effect of dose of bovine growth hormone on lactation of dairy cows.** J. Dairy Sci. 68, 1109-1115.

Eppard P.J., Bauman D.E., Bitman J., Wood D., Akers R.M. and House W.A. (1985b) **Effect of dose of bovine growth hormone on milk composition : a-lactalbumin, fatty acids and mineral elements.** J. Dairy Sci. 68, 3047-3054.

Eppard P.J., Bauman D.E., Curtis C.R., Erb H.N., Lanza G.M. and DeGeeter M.J. (1987) **Effect of 188 day treatment with somatotropin on health and reproductive performance of lactating dairy cows.** J. Dairy Sci. 70, 582-591.

Eppard P.J., Lanza G.M., Hudson S., Cole W.J., Hintz R.L., White T.C., Ribelin W.E., Hammond B.G., Bussen S.C., Leak R.K. and Metzger L.E. (1988) **Response of lactating dairy cows to multiple injections of sometribove, USAN (recombinant methionyl bovine somatotropin) in a prolonged release system. Part I. Production response.** J. Dairy Sci. 71 (Suppl. 1), 184.

Erb H.N. (1987) **Interrelationship among production and clinical disease in dairy cattle : a review.** Anim. Vet. J. 28, 326-329.

Escher J.T.M. and Van den Berg G. (1987) **Investigations on the influence of r-DNA bovine somatotropin on milk properties.** Report NOV-1245, Neth. Dairy Res., Ede.

Escher J.T.M. and Van den Berg G. (1988a) **The effect of somidobove in a slow release vehicle on some milk constituents and the renneting properties.** Report NOV-1309, Neth. Inst. Dairy Res., Ede.

Escher J.T.M. and Van den Berg G. (1988b) **The effect of somidobove in a sustained release vehicle on some milk constituents and cheesemaking properties.** Report NOV-1316, Neth. Inst. Dairy Res., Ede.

Esslemont R.J., Baile J.H. and Cooper M.J. (1985) "Infertility management in dairy cattle". Publishers Collins, London.

Etherton T.D. (1989) **The mechanisms by which porcine growth hormone improves pig growth performance.** In "Biotechnology in Growth Regulation" (Eds. R.B. Heap, C.G. Prosser and G.E. Lamming), Butterworth, London.

Etherton T.D. and Evock C.M. (1986) **Stimulation of lipogenesis in bovine adipose tissue by insulin and insulin-like growth factor.** J. Anim. Sci. 62, 357-362.

Etherton T.D., Evock C.M., Chung C.S., Walton P.E., Sillence M.N., Magri K.A. and Ivy R.E. (1986a) **Stimulation of pig growth perfomance by long term treatment with pituitary porcine growth hormone (pGH) and a recombinant (pGH).** J. Anim. Sci. 63 (Suppl. 1), 219.

Etherton T.D., Wiggins J.P., Chung C.S., Evock C.M., Redhun J.F. and Walton, P.E. (1986b) **Stimulation of pig growth performance by porcine growth hormone and growth hormone releasing factor.** J. Anim. Sci. 63, 1389-1399.

Etherton T.D., Evock C.M. and Kensinger R.S. (1987a) **Native and recombinant bovine growth hormone antagonize insulin action in cultured bovine adipose tissue.** Endocr. 121, 699-703.

Etherton T.D., Wiggins J.P., Evock C.M., Chung C.S., Rebhun J.F., Walton P.E. and Steele N.C. (1987b) **Stimulation of pig growth performance by porcine growth hormone : determination of the dose-response relationship.** J. Anim. Sci. 64, 433-443.

Ethier S.P., Kudla A. and Cundiff K.C. (1987) **Influence of hormone and growth factor interactions on the proliferative potential of normal rat mammary epithelial cells in vitro.** J. Cell. Physiol. 132, 161-167.

Evans F.D., Osborne V.R., Evans N.M., Morris J.J. and Hacker R.R. (1988) **Effect of different patterns of administration of recombinant porcine somatotropin (rpST) to pigs from 5 to 15 weeks of age.** J. Anim. Sci. 66 (Suppl. 1), 256.

Evans H.M. and Simpson M.E. (1931) **Hormones of the anterior hypophysis.** Am. J. Physiol. 98, 511-546.

Everett R.W. (1987) **How will BST affect dairy genetics in the 1990's?** Hoards Dairyman, April 10.

Evock C.M., Etherton T.D., Chung C.S. and Ivy R.E. (1988) **Pituitary porcine growth hormone (pGH) and a recombinant pGH analogue stimulate pig growth performance in a similar manner.** J. Anim. Sci. 66, 1928-1941.

Fabry J., Claes V. and Ruelle L. (1987) **Effect of growth hormone on heifer meat production.** Reprod. Nutr. Dévelop. 27, 591-600.

Farries E. and Profittlich C. (1987) **The influence of applicated bovine somatotropin on some metabolic criteria in dairy cows.** 38th Ann. Meeting Europ. Ass. Anim. Prod., Lisbon, Portugal, p. 432.

Farries E. and Profittlich C. (1988) **Veränderung einiger stofwechselparameter bei der Milchkuh. BST-Symposium.** Landbauforschung Völkenrode, Sonderheft, pp. 135-158.

Faverdin P., Hoden A. and Coulon J.B. (1987) **Recommandations alimentaires pour les vaches laitières.** Bull. Tech. CRZV Theix, INRA 70, 133-152.

Ferguson K.A. (1954) **Prolonged stimulation of wool growth following injections of ox growth hormone.** Nature (Lond.) 174, 411.

Fleet I.R., Fullerton F.M. and Mepham T.B. (1986) **Effects of exogenous growth hormone on mammary function in lactating Jersey cows.** J. Physiol. 376, 19 p.

Fleet I.R., Fullerton F.M., Heap R.B., Mepham T.B., Gluckman P.D. and Hart I.C. (1988) **Cardiovascular and metabolic responses during growth hormone treatment of lactating sheep.** J. Dairy Res. 55, 479-485.

Flint D.J. (1989) **Alternatives to growth hormone for the manipulation of animal performance.** In "Use of Somatotropin in Livestock Production" (Eds. K. Sejrsen, M. Vestergaard and A. Neimann-Srensen), Elsevier Applied Science, London and New York, pp. 51-60.

Flint D.J., Coggrave H., Futter C.E., Gardner M.J. and Clarke T.J. (1986) **Stimulatory and cytotoxic effects of an antiserum to adipocyte plasma membranes on adipose tissue metabolism in vitro and in vivo.** Int. J. Obesity 10, 69-77.

Francis G.L., Read C., Ballard F.J., Bagley C.J., Upton F.M., Gravestock M. and Wallace J.C. (1986) **Purification and partial sequence analysis of insulin-like growth factor-1 from bovine colostrum.** Biochem. J. 223, 207-213.

Francis G.L., Upton F.M., Ballard F.J., McNeil K.A. and Wallace J.C. (1988) **Insulin-like growth factors 1 and 2 in bovine colostrum.** Biochem. J. 251, 95-103.

Frangione T. and Cady R.A. (1988) **Effects of bovine somatotropin on sire summaries for milk production and milk yield heritabilities.** J. Dairy Sci. 71 (Suppl. 1), 239.

Fronk T.J., Peel C.J., Bauman D.E. and Gorewit R.C. (1983) **Comparison of different patterns of exogenous growth hormone administration on milk production in Holstein cows.** J. Anim. Sci. 57, 699-705.

Fullerton F.M., Mepham T.B., Fleet I.R. and Heap R.B. (1989) **Changes in mammary uptake of essential amino acids in lactating Jersey cows in response to exogenous bovine pituitary somatotropin.** In "Biotechnology in Growth Regulation" (Eds. R.B. Heap, C.G. Prosser and G.E. Lamming), Butterworth, London.

Furniss S.J., Stroud A.J., Brown A.C.G. and Smith C. (1988) **Milk production, feed intakes and weight change of autumn calving, flat rate fed dairy cows given two-weekly injections of recombinantly derived bovine somatotropin (BST).** Anim. Prod. 46, 483.

Futter C.E. and Flint D.J. (1987) **Long-term reduction of adiposity in rats after passive immunization with antibodies to rat fat cell plasma membranes.** In "Recent Advances in Obesity Research V" (Ed. E.M. Berry), John Libbey and Co., London, chapter 27.

Gardner M.J. and Flint D.J. (1988) **Effects of an antiserum to rat growth hormone (rGH) on growth and serum IGF-I levels in neonatal rats.** J. Endocr. 117 (Suppl.), 55.

Gertler A., Cohen N. and Maoz A. (1983) **Human growth hormone but not ovine or bovine growth hormone exhibit galactopoietic prolactin-like activity in organ culture from bovine lactating mammary gland.** Molec. Cell Endocr. 33, 169-182.

Giles D.D. (1942) **An experiment to determine the effect of growth hormone of the anterior lobe of the pituitary gland on swine.** Amer. J. Vet. Res. 3, 77-86.

Gluckman P.D. and Breier B.H. (1989) **The regulation of the growth hormone receptor.** In "Biotechnology in Growth Regulation" (Eds. R.B. Heap, C.G. Prosser and G.E. Lamming), Butterworth, London.

Gluckman P.D., Butler J.H. and Elliot T.B. (1983) **The ontogeny of somatotrophic binding sites in ovine hepatic membranes.** Endocr. 112, 1607-1612.

Gluckman P.D., Breier B.H. and Davis S.R. (1987) **Physiology of the somatotropic axis with particular reference to the ruminant.** J. Dairy Sci. 70, 442-466.

Goff J.P., Caperna T.J., Campbell R.G. and Steele N.C. (1988) **Interactions of porcine growth hormone (pGH) administration and dietary energy intake on circulating vitamin D metabolite concentrations in growing pigs.** J. Anim. Sci. 66 (Suppl. 1), 291.

Goodman G.T., Akers R.M., Friderici K.H. and Tucker H.A. (1983) **Hormonal regulation of a-lactalbumin secretion from bovine mammary tissue cultured in vitro.** Endocr. 112, 1324-1330.

Gopinath R. and Etherton T.D. (1988) **Pocine growth hormone (pGH) increases hepatic glucose production rate and impairs glucose clearance.** J. Anim. Sci. 66 (Suppl. 1), 292.

Graf L. and Li C.H. (1974) **Isolation and properties of two biologically active fragments from limited tryptic hydrolysis of bovine and ovine pituitary growth hormones.** Biochemistry 13, 5408-5415.

Gravert H.O. (1988) **Bendlung von Kühen mit gentechnologisch gewonnenem Wachstumshormon.** DMZ Welt der Milch 42, 917-918.

Gravert H.O. (1989) **Influences of somatotropin on evaluation of genetic merit for milk production.** In "Use of Somatotropin in Livestock Production" (Eds. K. Sejrsen, M. Vestergaard and A. Neimann-Srensen), Elsevier Applied Science, London and New York, pp. 120-131.

Grings E.E., De Avila D.M. and Reeves J.J. (1987) **Reproduction and growth in post-pubertal dairy heifers treated with recombinant somatotropin.** J. Anim. Sci. 65 (Suppl. 1), 248.

Hall K. and Sara V.R. (1984) **Somatomedin levels in childhood, adolescence and adult life.** Clin. Endocr. Metab. 13, 91-112.

Hancock D.L. and Preston R.L. (1988) **Titration of bovine somatotropin (bST) dose response which maximizes plasma urea nitrogen (PUN) depression in feedlot steers.** J. Anim. Sci. 66 (Suppl. 1), 254.

Hanrahan J.P., Quirke J.F., Bomann W., Allen P., McEwan J.C., Fitzsimons J.M., Kotzian J. and Roche J.F. (1986) **β-agonists and their effects on growth and carcass quality.** In "Recent Advances in Animal Nutrition" (Eds. W. Haresign and D.J.A. Cole), Butterworth, London, pp. 125-138.

Hanwell A. and Linzell J.L. (1973) **The time course of cardiovascular changes in lactation in the rat.** J. Physiol. 233, 93-109.

Hara K., Chen C.J.H. and Sonenberg M. (1978) **Recombination of the biologically active peptides from a tryptic digest of bovine growth hormone.** Biochemistry 17, 550-556.

Hard D.L., Cole W.J., Franson S.E., Samuels W.A., Bauman D.E., Erb H.N., Huber, J.T. and Lamb, R.C. (1988) **Effect of long term sometribove, USAN (recombinant methionyl bovine somatotropin) treatment in a prolonged release system on milk yield, animal health and reproductive performance-pooled across four sites.** J. Dairy Sci. 71 (Suppl. 1), 210.

Harper J.M.M., Soar J.B. and Buttery J.P. (1987) **Changes in protein metabolism of ovine primary muscle cultures on treatment with growth hormone, insulin, insulin-like growth factor I or epidermal growth factor.** J. Endocr. 112, 87-96.

Hart I.C. (1988) **Altering the efficiency of milk production of dairy cows with somatotropin.** In "Nutrition and Lactation in the Dairy Cow" (Ed. P.C. Garnsworthy), Butterworth, London, pp. 232-248.

Hart I.C. and Johnsson I.D. (1986) **Growth hormone and growth in meat producing animals.** In "Control and Manipulation of Animal Growth" (Eds. P.J. Buttery, D.B. Lindsay and N.B. Haynes), Butterworth, London, pp. 135-159.

Hart I.C., Bines J.A., Morant S.V. and Ridley J.L. (1978) **Endocrine control of the levels of hormones (prolactin, growth hormone, thyroxine and insulin) and metabolites in the plasma of high and low yielding cattle at various stages of lactation.** J. Endocr. 77, 333-345.

Hart, I.C., Bines J.A. and Morant S.V. (1979) **Endocrine control of energy metabolism in the cow : correlations of hormones and metabolites in high and low yielding cows for stages of lactation.** J. Dairy Sci. 62, 270-277.

Hart I.C., Bines J.A. and Morant S.V. (1980) **The secretion and metabolic clearance rates of growth hormone, insulin and prolactin in high and low yielding cattle at four stages of lactation.** Life Sci. 27, 1839.

Hart I.C., Chadwick P.M.E., Boone T.C., Langley K.E., Rudman C. and Souza L.M. (1984) **A comparison of the growth-promoting, lipolytic, diabetogenic and immunological properties of pituitary and recombinant-DNA-derived bovine growth hormone (somatotropin).** Biochem. J. 224, 93-100.

Hart I.C., Bines J.A. and Morant S.V. (1985a) **The effect of injecting or infusing low doses of bovine growth hormone on milk yield, milk composition and the quantity of hormone in the milk serum of cows.** Anim. Prod. 40, 243-250.

Hart I.C., Chadwick P.M.E., James S. and Simonds A.D. (1985b) **Effect of intravenous bovine growth hormone or human pancreatic growth hormone-releasing factor on milk production and plasma hormones and metabolites in sheep.** J. Endocr. 105, 189-196.

Hartnell G.E. (1986) **Evaluation of vitamins in milk produced from cows treated with placebo and CP115099 in a prolonged release system.** Monsanto Technical Report MSL-7031.

Heap R.B., Fleet I.R., Fullerton F.M., Davis A.J., Goode J.A., Hart I.C., Pendleton J.W., Prosser C.G., Sylvester L.M. and Mepham T.B. (1989) **A comparison of the mechanisms of action of bovine pituitary-derived and recombinant somatotropin (ST) in inducing galactopoiesis in the cow during late lactation.** In "Biotechnology in Growth Regulation" (Eds. R.B. Heap, C.G. Prosser and G.E. Lamming), Butterworth, London.

Heeschen W. (1988) **bST in der Milch. BST-Symposium.** Landbauforschung Völkenrode, Sonderheft, pp. 220-226.

Heird C.E., Hallford D.M., Spoon R.A., Holcombe D.W., Pope T.C., Olivares V.H. and Herring M.A. (1988) **Growth and hormone profiles in fine-wool ewe lambs after long-term treatment with ovine growth hormone.** J. Anim. Sci. 66 (Suppl. 1), 201.

Hemken R.W., Harmon R.J., Silvia W.J., Heersche G. and Eggert R.G. (1988) **Response of lactating dairy cows to a second year of recombinant bovine somatotropin (BST) when fed two energy concentrations.** J. Dairy Sci. 71 (Suppl. 1), 122.

Henricson B. and Ullberg S. (1960) **Effect of pig growth hormone on pigs.** J. Anim. Sci. 19, 1002-1008.

Hizuka N., Takano K., Asakawa K., Miyakawa M., Tanaka I., Horikawa R., Hasegawa S., Mikasa Y., Saito S., Shibasaki T. and Shizume K. (1987) **In vivo effects of insulin-like growth factor I in rats.** Endocr. Jpn. 34 (Suppl. 1), 115-121.

Holder A.T., Aston R., Preece M.A. and Ivanyi J. (1985) **Monoclonal antibody-mediated enhancement of growth hormone activity in vivo.** J. Endocr. 107, R9-12.

Honegger A. and Humbel R.E. (1986) **Insulin-like growth factors I and II in fetal and adult bovine serum.** J. Biol. Chem. 261, 569-575.

Huber J.T. (1987) **The production response of BST : feed additives, heat stress and injection intervals.** In "National Invitational Workshop on Bovine Somatotropin", St Louis, USA, pp. 57-60.

Huber, J.T., Franson, S.E., Hoffman R.G. and Hard D.L. (1988a) **Relationship of production level and days postpartum to response of cows to sometribove.** J. Dairy Sci. 71 (Suppl. 1), 207.

Huber J.T., Willman S., Marcus K. and Theurer C.B. (1988b) **Effect of sometribove (SB) USAN (recombinant methionyl bovine somatotropin) injected in lactating**

cows at 14-d intervals on milk yields, milk composition and health. J. Dairy Sci. 71 (Suppl. 1), 207.
Hughes J.P. (1979) **Identification and characterization of high and low affinity binding sites for growth hormone in rabbit liver.** Endocr. 105, 414-420.
Huisman J., van Weerden E.J., van der Hal W., Verstegen M.W.A., Kanis E. and van den Wal, P. (1988) **Effect of rpST treatment on rate of gain in protein and fat in two breeds of pigs and as crossbreds.** J. Anim. Sci. 66 (Suppl. 1), 254.
Hutchison C.F., Tomlinson J.E. and McGee W.H. (1986) **The effects of exogenous recombinant or pituitary extracted bovine growth hormone on performance of dairy cows.** J. Dairy Sci. 69 (Suppl. 1), 152.
Huth F.W. and Schutzbar W.V. (1987) **Einflus der Milchleistung auf die Zwischenkalbezeit.** Deutsche Schwarzbunte 11, 12-14.
Imagawa W., Spencer E.M., Larson L. and Nandi S. (1986) **Somatomedin C substitutes for insulin for the growth of mammary epithelial cells from normal virgin mice in serum-free collagen gel cell culture.** Endocr. 119, 2695-2699.
Irvin R. and Trenkle A. (1971) **Influence of age, breed and sex on plasma hormones in cattle.** J. Anim. Sci. 32, 292-295.
Isaksson O.G., Lindahl A., Nilsson A. and Isgaard J. (1987) **Mechanism of the stimulatory effect of growth hormone on longitudinal growth hormone.** Endocr. Rev. 8, 426-438.
Ivy R.E., Baldwin C.D., Wolfrom G.W. and Mouzin D.E. (1986a) **Effect of various levels of recombinant porcine growth hormone (rpGH) injected intramuscularly in barrows.** J. Anim. Sci. 63 (Suppl. 1), 218.
Ivy R.E., Wolfrom G.W. and Edwards C.K. (1986b) **Effects of growth hormone and RALGRO (zeranol) in finishing beef cattle.** J. Anim. Sci. 63 (Suppl. 1), 217-218.
Jaster E.H. and Wegner T.N. (1981) **Beta-adrenergic receptor involvement in lipolysis of dairy cattle subcutaneous adipose tissue during dry and lactating state.** J. Dairy Sci. 64, 1655-1663.
Jenness R. (1985) **Biochemical and nutritional aspects of milk and colostrum.** In "Lactation" (Ed. B.L. Larson), The Iowa State Univ. Press, Ames, pp. 164-197.
Jenny B.F., Ellers J.E., Tingle R.B., Moore M., Grimes L.W. and Rock D.W. (1988) **Responses of dairy cows to recombinant bovine somatotropin in a sustained release vehicle.** J. Dairy Sci. 71 (Suppl. 1), 209.
Johke T. and Hodate K. (1977) **Bovine serum prolactin, growth hormone and triiodothyronine levels during late pregnancy and early lactation.** Jap. J. Zootech. Sci. 48, 772-776.
Johnsson I.D., Hart I.C. and Butler-Hogg B.W. (1985) **The effects of exogenous bovine growth hormone and bromocriptine on growth, body development, fleece weight and plasma concentrations of growth hormone, insulin and prolactin in female lambs.** Anim. Prod. 41, 207-217.
Johnsson I.D., Hathorn D.J., Wilde R.M., Treacher T.T. and Butler-Hogg B.W. (1987) **The effect of dose and method of administration of biosynthetic bovine somatotropin on live-weight gain, carcass composition and wool growth in young lambs.** Anim. Prod. 44, 405-414.
Journet M. and Chilliard Y. (1985) **Influence de l'alimentation sur la composition du lait. I. Taux butyreux, facteurs généraux.** Bull. Tech. CRZV Theix, INRA 60, 13-23.
Kanis E., van der Hel W., Huisman J., Nieuwhof G.J., Politiek R.D., Verstegen M.W.A., van der Wal P. and van Weerden E.J. (1988) **Effect of recombinant somatotropin (rpST) treatment on carcass characteristics and organ weights of growing pigs.** J. Anim. Sci. 66 (Suppl. 1), 280.

Kann G., Périer A. and Martinet J. (1988) **Use of human growth hormone releasing factor (hGRF 1-29)NH₂ as a mammotropic hormone in the ewe.** J. Anim. Sci. 66 (Suppl. 1), 389.

Kaplan S.A. (1965) **Growth hormone.** Am. J. Dis. Child. 110, 232-238.

Karg H. (1987) **Hormonale Manipulation des Wachstums.** Übers. Tierernährung 15, 1-28.

Karg H. and Mayer H. (1987) **Manipulation der Laktation.** Übers. Tierernährung 15, 29-58.

Kazmer G.W., Barnes M.A., Akers R.M. and Pearson R.E. (1986) **Effect of genetic selection for milk yield and increased milking frequency on plasma growth hormone and prolactin concentration in Holstein cows.** J. Anim. Sci. 63, 1220-1227.

Kempster A.J. (1988) **Market requirements and their relation to carcass quality.** In "Beta-agonists and Their Effects on Animal Growth and Carcass Quality" (Ed. J.P. Hanrahan). A Seminar in the CEC programme of Coordination of Research in Animal Husbandry, Brussels, 1987. Elsevier Applied Science, London, pp. 72-82.

Kempster A.J., Cook G.L. and Grantley-Smith M. (1986) **National estimates of the body composition of British cattle, sheep and pigs with special reference to trends in fatness.** A review. Meat Sci. 17, 107-138.

Kik N. and Cook R.M. (1986) **Effects of bovine somatotropin and IsoPlus on milk production.** J. Dairy Sci. 69 (Suppl. 1), 158.

Kindstedt P.S., Rippe J.K. Pell A.N. and Hartnell G.F. (1988) **Effect of long-term administration of sometribove, USAN (recombinant methionyl bovine somatotropin) in a prolonged release formulation on protein distribution in Jersey milk.** J. Dairy Sci. 71 (suppl. 1), 96.

Kirchgessner M., Roth F.X., Schams D. and Karg H. (1987) **Influence of exogenous growth hormone (GH) on performance and plasma GH concentrations of female veal calves.** J. Anim. Physiol. Anim. Nutr. 58, 50-59.

Kirchgessner M., Schwab W. and Muller H.L. (1989) **Effect of bovine growth hormone on energy metabolism of lactating cows in long-term administration.** In "Energy Metabolism of Farm Animals" (Eds. Y. van der Honing and W.H. Close), Pudoc Wageningen, The Netherlands, pp. 143-146.

Knight C.D., Azain M.J., Kasser T.R., Sabacky M.J., Baile C.A., Buonomo F.C. and McLaughlin C.L. (1988) **Functionality of an implantable 6-week delivery system for porcine somatotropin (PST) in finishing hogs.** J. Anim. Sci. 66 (Suppl. 1), 257-258.

Knobil E. and Greep R.O. (1959) **The physiology of growth hormone with particular reference to its action in the Rhesus monkey and the "species specificity" problem.** In "Recent Progress in Hormone Research" (Ed. G. Pincus), Academic Press, New York, vol. 15, pp. 1-69.

Koprowski J.A. and Tucker H.A. (1973) **Bovine serum growth hormone, corticoids and insulin during lactation.** Endocr. 93, 645-651.

Kraft L.A., Haines D.R. and DeLay R.L. (1986) **The effect of daily injections of recombinant porcine growth hormone (rpGH) on growth, feed efficiency, carcass composition and selected metabolic and hormonal parameters in finishing swine.** J. Anim. Sci. 63 (Suppl. 1), 218.

Krivi G.G., Salsgiver W.J., Staten N.R., Hauser S.D., Rowold E., Kasser T.R., White T.C., Eppard P.J., Lanza G.M. and Wood D.C. (1988) **Identification of residues of somatotropin involved in receptor binding and biological activity.** In "70th Annual Meeting of the Endocrine Society", New Orleans, Abs. 257.

Kronfeld D.S. (1987) **Health risks in dairy cows given biosynthetic somatotropin.** Nutr. Inst. Proc., Nat. Feed Ingr. Ass., West Des Moines, Iowa.

Lakehal F., Crompton L.A. and Lomax M.A. (1989) **The effect of growth hormone on hind-limb muscle metabolism.** In "Biotechnology in Growth Regulation" (Eds. R.B. Heap, C.G. Prosser and G.E. Lamming), Butterworth, London.
Lamb R.C., Anderson M.J., Henderson S.L., Call J.W., Callan R.J., Hard D.L. and Kung L. (1988) **Production response of Holstein cows to sometribove USAN (recombinant methionyl bovine somatotropin) in a prolonged release system for one lactation.** J. Dairy Sci. 71 (Suppl. 1), 208.
Lanza G.M., Baile C.A. and Collier R.J. (1988a) **Development and potential of BST.** Nutr. Inst. Proc., Nat. Feed Ingr. Ass., West Des Moines, USA, 14p.
Lanza G.M., Eppard P.J., Miller M.A., Franson S.E., Ganguli S., Hintz R.L., Hammond B.G., Busser S.C., Leak R.K. and Metzger L.E. (1988b) **Response of lactating dairy cows to multiple injections of sometribove, USAN (recombinant methionyl bovine somatotropin) in a prolonged release system. Part III. Changes in circulating analytes.** J. Dairy Sci. 71 (Suppl. 1), 184.
Lanza G.M., White T.C., Dyer S.E., Hudson S., Franson S.E., Hintz R.L., Duque J.A., Bussen S.C., Leak R.K. and Metzger L.E. (1988c) **Response of lactating dairy cows to intramuscular or subcutaneous injection of sometribove, USAN (recombinant methionyl bovine somatotropin) in a 14-day prolonged release system. Part II. Changes in circulating analytes.** J. Dairy Sci. 71 (Suppl. 1), 195.
Lanza G.M., Krivi G.G., Bentle L.A., Eppard P.J., Kung L., Hintz R.L., Ryan R.L. and Miller M.A. (1988d) **Comparison of the galactopoietic activity of several recombinant bovine somatotropin variants and pituitary derived bovine somatotropin.** In "70th Annual Meeting of the Endocrine Society", New Orleans, Abs. 242.
Lapierre H., Pelletier G., Petitclerc D., Dubreuil P., Morisset J., Gaudreau P., Couture Y. and Brazeau P. (1988a) **Effect of human growth hormone-releasing factor (1-29)$NH_2$ on growth hormone release and milk production in dairy cows.** J. Dairy Sci. 71, 92-98.
Lapierre H., Petitclerc D., Pelletier G., Dubreuil P., Gaudreau P., Couture Y., Morisset J. and Brazeau P. (1988b) **Effects of growth hormone-releasing factor (GRF) and (or) thyrotropin-releasing factor (TRF) on energy and nutrient digestibilities and balances in male dairy calves.** J. Anim. Sci. 66 (Suppl. 1), 438.
Laron Z. (1983) **Deficiencies of growth hormone and somatomedins in man.** Special Topics Endocrin. Metab. 5, 149-199.
Lebenthal E. and Leung Y.-K. (1987) **The impact of development of the gut on infant nutrition.** Pediatr. Ann. 16, 211-222.
Lebenthal E., Lee P.C. and Heitlinger L.A. (1983) **Impact of development of the gastrointestinal tract on infant feeding.** J. Pediatr. 102, 1-9.
Lecce J.G. (1972) **Selective absorption of macromolecules into intestinal epithelium and blood by neonatal mice.** J. Nutr. 102, 69-75.
Lecce J.G. (1973) **Effect of dietary regimen on cessation of uptake of macromolecules by piglet intestinal epithelium (closure) and transport to the blood.** J. Nutr. 103, 751-756.
Lecce J.G. (1979) **Intestinal barriers to water-soluble macromolecules.** Environ. Health Perspect. 33, 57-60.
Lee M.O. and Schaffer N.K. (1934) **Anterior pituitary growth hormone and the composition of growth.** J. Nutr. 7, 377.
Leenanuruksa D. and McDowell G.H. (1987) **Diabetogenic effects of exogenous growth hormone demonstrated in vivo in sheep.** Proc. Nutr. Soc. Aust. 12, p. 175.

Leenanuruksa D. Smithard R., McDowell G.H., Gooden J.M., Jois M. and Niumsup P. (1985) **Effects of growth hormone on nutrient utilisation and blood flow in muscle tissue of growing calves.** Proc. Nutr. Soc. Aust. 10, p. 152.

Leenanuruksa D., Niumsup P., van der Walt J.G., Gooden J.M. and McDowell G.H. (1986) **Effect of exogenous growth hormone on hepatic-exchanges of glucose, free fatty acids and insulin in lactating ewes.** Proc. Nutr. Soc. Aust. 11, p. 95.

Lehninger A. (1970) "The Molecular Basis of Cell Structure and Function". Worth Publisher Co., New York.

Leitch, H.W., Burnside E.B., MacLeod G.K., McBride Kennedy, B.W. and Wilton J.W. (1987) **Genetic and phenotypic affects of administration of recombinant bovine somatotrophin to Holstein cows.** J. Dairy Sci. 70 (Suppl. 1), 128.

Lesniak M.A. and Roth J. (1976) **Regulation of receptor concentration by homologous hormone. Effect of human growth hormone on its receptor in IM-9 lymphocytes.** J. Biol. Chem. 251, 3720-3729.

Leung D.W., Spencer S.A., Cachianes G., Hammonds R.G., Collins C., Henzel W.J., Barnard R., Waters M.J., and Wood W.I. (1987) **Growth hormone receptor and serum binding protein : purification, cloning and expression.** Nature 330, 537-543.

Levinsky R.J. (1985) **Factors influencing intestinal uptake of food antigens.** Proc. Nutr. Soc. 44, 81-86.

Lewis K.J., Molan P.C., Bass J.J. and Gluckman P.D. (1988) **The lipolytic activity of low concentrations of insulin-like growth factors in ovine adipose tissue.** Endocr. 122, 2554-2557.

Linzell J.L. (1974) **Mammary blood flow and methods of identifying and measuring precursors of milk.** In "Lactation" (Eds. B.L. Larson and V.R. Smith), Academic Press, New York, vol. 1, pp. 143-225.

Lönnroth P., Assmundsson K., Edén S., Enberg G., Gouse I., Hall K. and Smith U. (1987) **Regulation of insulin-like growth factor II receptors by growth hormone and insulin in rat adipocytes.** Proc. Natl. Acad. Sci. USA 84, 3619-3622.

Lossouarn J. (1988) **Etude d'une formulation retard de zinc-methionyl bovine somatotropine pour la production laitière.** Compte-rendu d'essai. INA Paris, Grignon, Monsanto, France.

Lough D.S., Muller L.D., Kensinger R.S., Sweeney T.F. and Griel Jr. L.C. (1988) **Effect of added dietary fat and bovine somatotropin on the performance and metabolism of lactating dairy cows.** J. Dairy Sci. 71, 1161-1169.

Lund-Larsen T.R., Sundby A., Kruse V. and Velle W. (1977) **Relation between growth rate, serum somatomedin and plasma testosterone in young bulls.** J. Anim. Sci. 44, 189-194.

Lynch J.M., Barbano D.M., Bauman D.E. and Hartnell G.F. (1988a) **Influence of sometribove (recombinant methionyl bovine somatotropin) on the protein and fatty acid composition of milk.** J. Dairy Sci. 71 (Suppl. 1), 100.

Lynch J.M., Senyk G.F., Barbano D.M., Bauman D.E., and Hartnell G.F. (1988b) **Influence of sometribove (recombinant methionyl bovine somatotropin) on milk lipase and protease activity.** J. Dairy Sci. 71 (Suppl. 1), 100.

Machlin L.J. (1972) **Effect of porcine growth hormone on growth and carcass composition of the pig.** J. Anim. Sci. 35, 794-800.

Maes M., Underwood L.E., Gerard G. and Ketelslegers J.-M. (1984) **Relationship between plasma somatomedin-C and liver somatogenic binding site in neonatal rats during malnutrition and after short and long term refeeding.** Endocr. 115, 786-792.

Magri K.A., Gopinath R. and Etherton T.D. (1987) **Inhibition of lipogenic enzyme activities by porcine growth hormone (pGH)** J. Dairy Sci. 65 (Suppl. 1), 258.

Maiter D., Maes M., Underwood L.E., Fliesen T., Gerard G. and Ketelslegers J.-M. (1988) **Early changes in serum concentrations of somatomedin-C induced by dietary deprivation in rats : contributions of growth hormone receptor and post-receptor defects.** J. Endocr. 118, 113-120.

Maltin C.A., Hay S.M., Delday M.I., Smith F.G., Lobley G.E. and Reeds P.J. (1987) **Clenbuterol, a beta agonist, induces growth in innervated and denervated rat soleus muscle via apparently different mechanisms.** Biosci. Rep. 7, 525-532.

Malvern P.V., Head H.H., Collier R.J. and Buonomo F.C. (1987) **Periparturient changes in secretion and mammary uptake of insulin and in concentrations of insulin and insulin-like growth factors in milk of dairy cows.** J. Dairy Sci. 70, 2254-2265.

Marsh W.E., Galligan D.T. and Chalupa W. (1987) **Making economic sense of bovine somatotropin use in individual dairy herds.** In "Nutrient Partitioning", Am. Cyanamid Techn. Symp., California, USA, pp. 59-79.

Massagué J. and Czech M.P. (1982) **The subunit structures of two distinct receptors for insulin-like growth factors I and II and their relationship to the insulin receptor.** J. Biol. Chem. 257, 5038-5045.

McBride B.W., Burton J.L. and Burton J.H. (1988) **The influence of bovine growth hormone (somatotropin) on animals and their products.** Res. Dev. Agric. 5, 1-21.

McCusker R.H. and Campion D.R. (1986) **Effect of growth hormone-secreting tumors on body composition and feed intake in young female Wistar-Furth rats.** J. Anim. Sci. 63, 1126-1133.

McCutcheon S.N. and Bauman D.E. (1986) **Effect of chronic growth hormone treatment on responses to epinephrine and thyrotropin-releasing hormone in lactating cows.** J. Dairy Sci. 69, 44-51.

McDaniel B.T. and Hayes P.W. (1988). **Absence of interaction of merit for milk with recombinant bovine somatotropin.** J. Dairy Sci. 71 (Suppl. 1), 240.

McDowell G.H., Gooden J.M., van der Walt J.G., Smithard R., Leenanuruksa D. and Niumsup P. (1985) **Metabolism of glucose and free fatty acids in lactating ewes treated with growth hormone.** Proc. Nutr. Soc. Aust. 10, p. 155.

McDowell G.H., Hart I.C. and Kirby A.C. (1987a) **Local intra-arterial infusion of growth hormone into the mammary glands of sheep and goats : effects on milk yield and composition, plasma hormones and metabolites.** Aust. J. Biol. Sci. 40, 181-189.

McDowell G.H., Gooden J.M., Leenanuruksa M.J. and English A.W. (1987b) **Effects of exogenous growth hormone on milk production and nutrient uptake by muscle and mammary tissues of dairy cows in mid-lactation.** Aust. J. Biol. Sci. 40, 295-306.

McDowell G.H., Hart I.C., Bines J.A., Lindsay D.B. and Kirby A.C. (1987c) **Effects of pituitary-derived bovine growth hormone on production parameters and biokinetics of key metabolites in lactating dairy cows at peak and mid-lactation.** Aust. J. Biol. Sci. 40, 191-202.

McGuffey R.K., Green H.B. and Ferguson T.H. (1987a) **Lactation performance of dairy cows receiving recombinant bovine somatotropin by daily injection or in a sustained release vehicle.** J. Dairy Sci. 70 (Suppl. 1), 176.

McGuffey R.K., Green H.B. and Basson R.P. (1987b) **Performance of Holsteins given bovine somatotropin in a sustained delivery vehicle. Effect of dose and frequency of administration.** J. Dairy Sci. 70 (Suppl. 1), 177.

McGuffey R.K., Green H.B. and Basson R.P. (1988) **Protein nutrition of the somatotropin-treated cow in early lactation.** J. Dairy Sci. 71 (Suppl. 1), 120.

McLaren D.G., Grebner G.L., Bechtel P.J., McKeith F.K., Novakofski J.E., Easter R.A., Jones R.W. and Dalrymple R.H. (1987) **Effect of graded levels of natural porcine** somatotropin (PST) on growth performance of 57 to 103 kg pigs. J. Anim. Sci. 65 (Suppl. 1), 245.

McLaughin C.L., Baile C.A., Qui S.-Z. and Wang L.-C. (1988) **Responses of Beijing Black hogs to porcine somatotropin (PST) treatment.** J. Anim. Sci. 66 (Suppl. 1), 255.

McNeish A.S. (1984) **Enzymatic maturation of the gastrointestinal tract and its relevance to food allergy and intolerance in infancy.** Ann. Allergy 53, 643-648.

McPherson J.M. and Livingston D.J. (1989) **Protein engineering : new approaches to improved therapeutic proteins.** Part III. Pharmaceut. Tech. 13, 33-42.

McShane T.M., Schillo K.K., Boling J.A., Bradley N.W. and Hall J.B. (1988) **Effects of somatotropin and dietary energy on development of beef heifers. I. Growth and puberty.** J. Anim. Sci. 66 (Suppl. 1), 252-253.

Mepham T.B., Lawrence S.E., Peters A.R. and Hart I.C. (1984) **Effects of growth hormone on mammary function in lactating goats.** Horm. Metab. Res. 16, 248-253.

Merimee T.J., Zapf J. and Froesch E.R. (1982) **Insulin-like growth factors in the fed and fasted states.** J. Clin. Endocr. Metab. 55, 999-1002.

Meulig C., Zapf J. and Froesch E.R. (1978) **NSILA-carrier protein abolishes the action of non-suppressible insulin-like activity (NSILA-S) on perfused rat heart.** Diabetologia 14, 253-259.

Meyer R.M., McGuffey R.K., Basson R.P., Rakes A.H., Harrison J.H., Emery R.S., Muller L.D. and Block E. (1988) **The effect of somidobove sustained release injection on the lactation performance of dairy cattle.** J. Dairy Sci. 71 (Suppl. 1), 207.

Miki N., Onu M. and Skizume K. (1984) **Evidence the opiatergic and a-adrenergic mechanisms stimulate rat growth hormone release via growth hormone releasing factor** (GRF) Endocr. 114, 1950-1952.

Mills J.B., Howard S.C., Scapa S. and Wilhelmi A.E. (1970) **Cyanogen bromide cleavage and partial amino acid sequence of porcine growth hormone.** J. Biol. Chem. 245, 3407-3415.

Mollett T.A., DeGeeter M.J., Belyea R.L., Youngquist R.A. and Lanza G.M. (1986) **Biosynthetic or pituitary extracted bovine growth hormone induced galactopoiesis in dairy cows.** J. Dairy Sci. 69 (Suppl. 1), 118.

Moore J.H. and Christie W.W. (1979) **Lipid metabolism in the mammary gland of ruminant animals.** Prog. Lipid Res. 17, 347-395.

Moore W.V., Draper S. and Hung C.H. (1985) **Species variation in the binding of hGH to hepatic membranes.** Horm. Res. 21, 33-45.

Moore J.A., Rudman C.G., MacLachlan N.J., Fuller G.B., Burnett B. and Frane J.W. (1988) **Equivalent potency and pharmacokinetics of recombinant human growth hormones with or without an N-terminal methionine.** Endocr. 122, 2920-2926.

Moseley W.M., Krabill L.F. and Olsen R.F. (1982) **Effect of bovine growth hormone administered in various patterns on nitrogen metabolism in the Holstein steer.** J. Anim. Sci. 55, 1062-1070.

Moseley W.M., Krabill L.F., Friedman A.R. and Olsen R.F. (1985) **Administration of synthetic human pancreatic growth hormone-releasing factor for five days sustains raised serum concentrations of growth hormone in steers.** J. Endocr. 104, 433-439.

Moseley W.M., Huisman J. and Van Weerden E.J. (1987) **Serum growth hormone and nitrogen metabolism responses in young bull calves infused with growth hormone-releasing factor for 20 days.** Dom. Anim. End. 4, 51-59.

Muir L.A., Wien S., Duquette P.F., Rickes A.L. and Cordes E.H. (1983) **Effects of exogenous growth hormone and diethylstibestrol on growth and carcass composition of growing lambs.** J. Anim. Sci. 56, 1315-1323.

Munneke R.L., Sommerfeldt J.L. and Ludens E.A. (1988) **Lactational responses of dairy cows to recombinant bovine somatotropin.** J. Dairy Sci. 71 (Suppl. 1), 206.

Murphy L.J., Vrhovsek E. and Lazarus L. (1983) **Identification and characterization of specific growth hormone receptors in cultured human fibroblasts.** J. Clin. Endocr. Metab. 57, 1117-1124.

Murphy L.J., Bell G.I. and Friesen H.G. (1987) **Tissue distribution of insulin-like growth factor I and II messenger-ribonucleic acid in the adult rat.** Endocr. 120, 1279-1282.

Nadler A.C., Sonenberg M., New M.I. and Free C.A. (1967) **Growth hormone activity in man with components of tryptic digests of bovine growth hormone.** Metab. 16, 830-845.

Newcomb M.D., Grebner G.L., Bechtel P.J., McKeith F.K., Novakofski J., McLaren D.G., Easter R.A. and Jones R.W. (1988) **Response of 60 to 100 kg pigs treated with porcine somatotropin to different levels of dietary crude protein.** J. Anim. Sci. 66 (Suppl. 1), 281.

Nilsson A., Isgaard J., Lindahl A., Peterson L. and Isaksson O. (1987) **Effects of unilateral arterial infusion of GH and IGF-I on tibial longitudinal bone growth in hypophysectomized rats.** Calcif. Tiss. Int. 40, 91-96.

Nissley S.P. and Rechler M.M. (1984) **Somatomedin/insulin-like growth factor tissue receptors.** Clin. Endocr. Metab. 13, 43-67.

Nissley S.P. and Rechler M.M. (1985) **Insulin-like growth factors : biosynthesis, receptors and carrier proteins.** Horm. Prot. Pept. 12, 128-203.

Niumsup P., McDowell G.H., Leenanuruksa D. and Gooden J.M. (1985) **Plasma triglyceride metabolism in lactating ewes and growing calvers treated with growth hormone.** Proc. Nutr. Soc. Aust. 10, p. 154.

Novakofski J., Brenner K., Easter R., McLaren D., Jones R., Ingle D. and Bechtel P. (1988) **Effects of porcine somatotropin on swine metabolism.** Fed. Proc. Amer. Soc. Exptl. Biol. 2, p. 848.

Nytes A.J., Combs D.K. and Shook G.E. (1988) **Efficacy of recombinant bovine somatotropin injected at three dosage levels in lactating dairy cows of different genetic potentials.** J. Dairy Sci. 71 (Suppl. 1), 123.

Oldenbroek J.K., Garssen G.J., Forbes A.B. and Jonker L.J. (1987) **The effect of treatment of dairy cows of different breeds with recombinantly derived bovine somatotropin in a sustained delivery vehicle.** 38th Ann. Meeting Europ. Ass. Anim. Prod., Lisbon, Portugal, 32 p.

Ooi G.T. and Herington A.C. (1988) **The biological and structural characterization of specific serum binding proteins for the insulin-like growth factors.** J. Endocr. 118, 7-18.

Pabst K., Prokopek D., Peters K.H. and Krusch U. (1987a) **Effect of application of BST on technological properties of milk and quality of products.** Fed. Dairy Res. Centre, Kiel, FRG, Ann. Rep. B8.

Pabst, K., Roos N. and Sick H. (1987b) **Influence of BST-treatment of cows on milk protein pattern.** Fed. Dairy Res. Centre, Kiel, FRG, Ann. Rep. B9.

Paladini A.C., Peña C. and Poskus E. (1983) **Molecular biology of growth hormone.** CRC Crit. Rev. Biochem. 15, 25-56.

Palmiter R.D., Brinster R.L., Hammer R.E., Trumbauer M.E., Rosenfeld M.G., Birnberg N.C. and Evans R.M. (1982) **Dramatic growth of mice that develop from eggs** microinjected with metallothionein-growth hormone fusion genes. Nature 300, 611-615.

Palmquist D.L. (1988) **Response of high-producing cows given daily injections of recombinant bovine somatotropin from D 30-296 of lactation.** J. Dairy Sci. 71 (Suppl. 1), 206.

Pastoreau P., Barenton B., Blanchard M., Boivin G., Charrier J., Dulor J.-P. and Theriez M. (1988) **Effects of GRF 1-29 in normal and hypotrophic lambs.** Reprod. Nutr. Dévelop. 28, 253-256.

Peel C.J. (1988) **Bovine somatotropin (BST). A review of efficacy and mechanism of action in dairy cows.** European Association of Veterinary Pharmacology and Toxicology. Budapest, September 1st.

Peel C.J. and Bauman D.E. (1987) **Somatotropin and lactation.** J. Dairy Sci. 70, 474-486.

Peel C.J. and Bauman D.E., Gorewit R.C. and Sniffen C.J. (1981) **Effect of exogenous growth hormone on lactational performance in high yielding dairy cows.** J. Nutr. 111, 1662-1671.

Peel C.J., Fronk T.J., Bauman D.E. and Gorewit R.C. (1982) **Lactational response to exogenous growth hormone and abomasal infusion of a glucose-sodium caseinate mixture in high-yielding dairy cows.** J. Nutr. 112, 1770-1778.

Peel C.J., Fronk T.J., Bauman D.E. and Gorewit R.C. (1983) **Effect of exogenous growth hormone in early and late lactation on lactational performance in dairy cows.** J. Dairy Sci. 66, 776-782.

Peel C.J., Sandles L.D., Quelch K.J. and Herington A.C. (1985) **The effects of long-term administration of bovine growth hormone on the lactational performance of identical-twin dairy cows.** Anim. Prod. 41, 135-142.

Peel C.J., de Kerchove G., Schockmel L.R. and Craven N. (1988) **Recent developments in use of somatotropins in lactating dairy cows.** Proc. VI World Conf. Animal Prod., Helsinki, p. 391.

Peel C.J., Eppard P.J. and Hard D.L. (1989) **Evaluation of sometribove (methionyl bovine somatotropin) in toxicology and clinical trials in Europe and the United States.** In "Biotechnology in Growth Regulation" (Eds. R.B. Heap, C.G. Prosser and G.E. Lamming), Butterworth, London.

Pell A.N., Tsang D.S., Huyler M.T., Howlett B.A. and Kunkel J. (1988) **Responses of Jersey cows to treatment with sometribove, USAN (recombinant methionyl bovine somatotropin) in a prolonged release system.** J. Dairy Sci. 71 (Suppl. 1), 206.

Pell J.M. and Bates P.C. (1987) **Collagen and non-collagen protein turnover in skeletal muscle of growth hormone-treated lambs.** J. Endocr. 115, R1-R4.

Perdue J.F. (1984) **Chemistry, structure, and function of insulin-like growth factors and their receptors : a review.** Can. J. Biochem. Cell Biol. 62, 1237-1245.

Peters J.P. (1986) **Consequences of accelerated gain and growth hormone administration for lipid metabolism in growing beef steers.** J. Nutr. 116, 2490-2503.

Phillips L.S., Fusco A.C., Unterman T.G. and del Greco F. (1984) **Somatomedin inhibitor in uremia.** J. Clin. Endocr. Metab. 59, 764-772.

Phipps R.H. (1987) **The use of prolonged release bovine somatotropin in milk production.** Int. Dairy Fed. Congr., Helsinki, Finland, 23 p.

Phipps R.H. (1988) **The use of prolonged release somatotropin in milk production.** Int. Dairy Fed. Bull. No. 228.

Phipps R.H. (1989) **A review of the influence of somatotropin on health, reproduction and welfare in lactating dairy cows.** In "Use of Somatotropin in Livestock Production" (Eds. K. Sejrsen, M. Vestergaard and A. Neimann-Srensen), Elsevier Applied Science, London and New York, pp. 88-119.

Phipps R.H. and Weller R.F. (1988) **Efficacy data for British Friesian treated for two consecutive lactations with a prolonged release formulation of bovine somatotropin.** XV World Buiatrics Conference, Palma, Oct. 11-14.

Phipps R.H., Weller R.F., Austin A.R., Craven N. and Peel C.J. (1988) **A preliminary report on a prolonged release formulation of bovine somatotropin with particular reference to animal health.** Vet. Rec. 122 : 512-513.

Plouzek C.A., Vale W., Rivier J., Anderson L.L. and Trenkle A. (1988) **Growth hormone-releasing factor on growth hormone secretion in prepubertal calves.** Proc. Soc. Exp. Biol. Med. 188, 198-205.

Pocius P.A. and Herbein J.H. (1986) **Effects of in vivo administration of growth hormone on milk production and in vitro hepatic metabolism in dairy cattle.** J. Dairy Sci. 69, 713-720.

Prosser C.G. and Mepham T.B. (1989) **Mechanism of action of bovine somatotropin in increasing milk secretion in dairy ruminants.** In "Use of Somatotropin in Livestock Production" (Eds. K. Sejrsen, M. Vestergaard and A. Neimann-Srensen), Elsevier Applied Science, London and New York, pp. 1-17.

Prosser C.G., Fleet I.R., Hart I.C. and Heap R.B. (1987a) **Changes in concentrations of insulin-like growth factor I (IGF-I) in milk during bovine growth hormone treatment in the goat.** J. Endocr. 112 (Suppl.), 65.

Prosser C.G., Davis A.J., Fleet I.R., Rees L.H. and Heap R.B. (1987b) **Mechanism of transfer of IGF-I into milk.** J. Endocr. 115 (Suppl.), 91.

Prosser C.G., Sankaran L., Hennighausen L. and Topper Y.J. (1987c) **Comparison of the roles of insulin-like growth factor I in casein gene expression and in the development of a-lactalbumin and glucose transport activities in the mouse mammary epithelial cell.** Endocr. 120, 1411-1416.

Prosser C.G., Fleet I.R., Corps A.N., Heap R.B. and Froesch E.R. (1988) **Increased milk secretion and mammary blood flow during close-arterial infusion of insulin-like growth factor I (IGF-I) into the mammary gland of the goat.** J. Endocr. 117 (Suppl.), 248.

Prosser C.G., Fleet I.R. and Heap R.B. (1989) **Action of IGF-1 on mammary function.** In "Biotechnology in Growth Regulation" (Eds. R.B. Heap, C.G. Prosser and G.E. Lamming), Buttersworth, London.

Pullar R.A., Johnsson I.D. and Chadwick P.M.C. (1986) **Recombinant bovine somatotropin is growth promoting and lipolytic in fattening lambs.** Anim. Prod. 42, 433-434.

Purchas R.W., Macmillian K.L. and Hafs H.D. (1970) **Pituitary and plasma growth hormone levels in bulls from birth to one year of age.** J. Anim. Sci. 31, 358-363.

Pursel V.G., Campbell R.C., Miller K.F., Behringer R.R., Palmiter R.D. and Brinster R.L. (1988) **Growth potential of transgenic pigs expressing a bovine growth hormone gene.** J. Anim. Sci. 66 (Suppl. 1), 267.

Quirke J.F., Kennedy L.G., Roche J.F., Hart I., Sheehan W., Coert A. and Allen P. (1985) **Responses of finishing steers to exogenous growth hormone and oestradiol.** Ir. Grassland and Anim. Prod. Assoc., 11th Ann. Res. Mtg.

Raben M.S. (1959) **Human growth hormone.** In "Recent Progress in Hormone Research" (Ed. G. Pincus), Academic Press, New York, pp. 71-114.

Rebhun J.F., Etherton T.D., Wiggins J.P., Chung C.S., Walton P.E. and Steek N. (1985) **Stimulation of swine growth performance by porcine growth hormone (pGH). Determination of the maximally effective pGH dose.** J. Anim. Sci. 61 (Suppl. 1), 251.

Rechler M.M. (1988) **Molecular insights into insulin-like growth factor biology.** J. Anim. Sci. 66 (Suppl. 3), 76-83.

Reklewska B. (1974) **A note on the effect of bovine somatotrophic hormone on wool production in growing lambs.** Anim. Prod. 19, 253-255.

Reinhardt C. (1984) **Macromolecular absorption of food antigens in health and disease.** Ann. Allergy 53, 597-601.

Remond B. (1985) **Influence de l'alimentation sur la composition du lait de vache- 2-taux protéique : facteurs généraux.** Bull. Tech. CRZV Theix, INRA, 62, 53-67.

Reynaert R., Marcus S. and Peeters P. (1976) **Influences of stress, age, sex on serum growth hormone and free fatty acid levels in cattle.** Horm. Metab. Res. 8, 109-114.

Rijpkema Y.S., van Reeuwijk L., Peel C.J. and Mol E.P. (1987) **Responses of dairy cows to long-term treatment with somatotropin in a prolonged release formulation.** Eur. Assoc. Anim. Prod., Lisbon, Portugal.

Ringuet H., Petitclerc D., Sorenson M., Gaudreau P., Pelletier G., Morisset J., Couture Y. and Brazeau P. (1988) **Effects of human somatocrinin (1-29) NH$_2$ (GRF) and photoperiod on carcass parameters and mammary growth of dairy heifers.** J. Dairy Sci. 71 (Suppl. 1), 193.

Roberton D.M., Paganelli R., Dinwiddie R. and Levinsky R.J. (1982) **Milk antigen absorption in the preterm and term neonate.** Arch. Dis. Child. 57, 369-372.

Roche J.F. and Quirke J.F. (1986) **The effects of steroid hormones and xenobiotics on growth of farm animals.** In "Control and Manipulation of Animal Growth" (Eds. P.J. Buttery, D.B. Lindsay and N.B. Haynes), Butterworth, London, pp. 39-51.

Roe J.A., Heywood C.M., Harper J.M.M. and Buttery P.J (1989) **Comparative aspects of selected hormones and growth factors on protein metabolism in adult and foetal ovine primary muscle cultures.** In "Biotechnology in Growth Regulation" (Eds. R.B. Heap, C.G. Prosser and G.E. Lamming), Butterworth, London.

Ronge H. and Blum J.W. (1988) **Somatomedin C and other hormones in dairy cows around parturition, in newborn calves and in milk.** J. Anim. Physiol. Anim. Nutr. 60, 168-176.

Ronge H., Blum J.W., Clement C., Jans F., Leuenberger H. and Binder H. (1988) **Somatomedin C in dairy cows related to energy and protein supply and to milk production.** Anim. Prod. 47, 165-183.

Roth R.A. (1988) **Structure of the receptor for insulin-like growth factor II : the puzzle amplified.** Science 239, 1269-1271.

Rowe-Bechtel C.L, Muller L.D., Deaver D.R. and Griel L.C. (1988) **Administration of recombinant bovine somatotropin (rbSt) to lactating dairy cows beginning at 35 and 70 days postpartum. I. Production response.** J. Dairy Sci. 71 (Suppl. 1), 166.

Salmon W.D. and Daughaday W.H. (1957) **A hormonally controlled serum factor which stimulates sulphate incorporation by cartilage in vitro.** J. Lab. Clin. Med. 49, 825-829.

Samuels W.A., Hard D.L., Hintz R.L., Olsson P.K., Cole W.J. and Hartnell G.F. (1988) **Long term evaluation of sometribove, USAN (recombinant methionyl bovine somatotropin) treatment in a prolonged release system for lactating cows.** J. Dairy Sci. 71 (Suppl. 1), 209.

Sandles L.D. and Peel C.J. (1987) **Growth and carcass composition of pre-pubertal dairy heifers treated with bovine growth hormone.** Anim. Prod. 44, 21-27.

Sartor O., Bowers C.Y., Reynolds G.A. and Momany F.A. (1985) **Variables determining the growth hormone response of His-D-Trp-Ala-Trp-D-Phe-Lys-NH$_2$ in the rat.** Endocr. 117, 1441-1447.

Schams D. (1987) **Analytik Wachstumshormon. Analytik des endogenen und exogenen Wachstumshormons beim Rind.** Süddeutsche Vers. Forsch. Anst. Milchwirtsch. Weihenstephan, Techn. Universität, München, Wissensch. Jahresbericht.

Schams D. (1989) **Somatotropin and related peptides in milk.** In "Use of Somatotropin in Livestock Production" (Eds. K. Sejrsen, M. Vestergaard and A. Neimann-Srensen), Elsevier Applied Science, London and New York, pp. 192-200.

Schams D., Winkler U., Schallenberger E. and Karg H. (1988) **Wachstumshormon und insulin like growth factor I (Somatomedin C) - Blutspiegel bei Rindern von der Geburt bis nach der pubertät.** Dtsch. Tierärztl. Wschr. 95, 353-408.

Schams D., Winkler U., Theyerl-Abele M. and Prokopp A. (1989) **Variation of BST and IGF-I concentrations in blood plasma of cattle.** In "Use of Somatotropin in Livestock Production" (Eds. K. Sejrsen, M. Vestergaard and A. Neimann-Srensen), Elsevier Applied Science, London and New York, pp. 18-30.

Schatz H., Katsilambros N., Hinz M., Voigt K.H., Nierle C. and Pfeiffer E.F. (1973) **Hypophysis and function of pancreatic islets. II. The effect of substitution with growth hormone and corticotrophin on insulin secretion and biosynthesis of proinsulin and insulin in isolated pancreatic islets of hypophysectomized rats.** Diabetologia 9, 140-144.

Schemm S.R. and Deaver D.R. (1988) **Administration of recombinant bovine somatotropin (rbst) to lactating dairy cows beginning at 35 and 70 days postpartum. IV. Effect on pituitary function and plasma estradiol.** J. Dairy Sci. 71 (Suppl. 1), 167.

Schneider P.L., Vecchiarelli B. and Chalupa W. (1987) **Bovine somatotropin and ruminally inert fat in early lactation.** J. Dairy Sci. 70 (Suppl. 1), 177.

Schoenle E., Zapf J., Humbel R.E. and Froesch E.R. (1982) **Insulin-like growth factor I stimulates growth in hypophysectomized rats.** Nature 296, 252-253.

Schoenle E., Zapf J., Hauri C., Steiner T. and Froesch E.R. (1985) **Comparison of in vivo effects of insulin-like growth factors I and II and of growth hormone in hypophysectomized rats.** Acta Endocr. 108, 167-174.

Schreiber A.B., Courgud P.D., Andre C.L., Vray B. and Strosberg A.D. (1980) **Antialprenolol anti-idiotypic antibodies bind to b-adrenergic receptors and modulate catecholamine sensitive adenylate cyclase.** Proc. Natl. Acad. Sci. USA 77, 7385-7389.

Schwartz J., Foster C.M. and Satin M.S. (1985) **Growth hormone and insulin-like growth factors I and II produce distinct alterations in glucose metabolism in 3T3-F442A adipocytes.** Proc. Natl. Acad. Sci. USA 82, 8724-8728.

Sechen S.J., McCutcheon S.N. and Bauman D.E. (1985) **Response to metabolic challenges in lactating dairy cows during short-term bovine growth hormone treatment.** J. Dairy Sci. 68 (Suppl. 1), 170.

Sechen S.J., Dunshea F.R. and Bauman D.E. (1988) **Mechanism of bovine somatotropin (bST) in lactating cows: effect on response to homeostatic signals (epinephrine and insulin).** J. Dairy Sci. 71 (Suppl. 1), 168.

Sechen S.J., Bauman D.E., Tyrrell H.F. and Reynolds P.J. (1989) **Effect of somatotropin on kinetics of nonesterified fatty acids and partition of energy, carbon, and nitrogen in lactating dairy cows.** J. Dairy Sci. 72, 59-67.

Sege K. and Peterson P.A. (1978) **Use of anti-idiotypic antibodies as cell surface receptor probes.** Proc. Natl. Acad. Sci. USA 75, 2443-2447.

Sejrsen K., Foldager J., Klastrup S. and Bauman D.E. (1986a) **Effect and mode of action of exogenous growth hormone on body tissue growth in heifers.** 37th Ann. Mtg. EAAP 1, 467.

Scjrscn K., Foldager J., Sorensen M.T., Akers R.M. and Bauman D.E. (1986b) **Effect of exogenous bovine somatotropin on pubertal mammary development in heifers.** J. Dairy Sci. 69, 1528-1535.

Shamay A., Cohen N., Niwa M. and Gertler A. (1988) **Effect of insulin-like growth factor I on deoxyribonucleic acid synthesis and galactopoiesis in bovine undifferentiated and lactating mammary tissue in vitro.** Endocr. 123, 804-809.
Shea B.T., Hammer R.E. and Brinster R.L. (1987) **Growth allometry of the organs in giant transgenic mice.** Endocr. 121, 1924-1930.
Simianer H. and Wollny C. (1988) **Impact of the potential use of bovine somatotropin on the accuracy of sire selection.** Polycopy, 21 p.
Skarda J., Urbanova E., Becka S., Houdebine L.M., Delouis C., Pichova D., Picha J. and Pilek J. (1982) **Effect of bovine growth hormone on development of goat mammary tissue in organ culture.** Endocr. Expt. 16, 19-31.
Skarda J., Krejci P., Slaba J. and Husakova E. (1989) **Blood constituents and subcutaneous adipose tissue metabolism of dairy cows after administration of recombinant bovine somatotropin (bST) in a prolonged release formulation.** In "Biotechnology in Growth Regulation" (Eds. R.B. Heap, C.G. Prosser and G.E. Lamming), Butterworth, London.
Smith R.D., Hansel W. and Coppock C.E. (1976) **Plasma growth hormone and insulin during early lactation in cows fed silage based diets.** J. Dairy Sci. 59, 248-254.
Soderholm C.G., Otterby D.E., Linn J.G., Wheaton J.E., Hansem W.P. and Annexstad R.J. (1986) **Effects of different doses of recombinant bovine somatotropin on circulating metabolites, hormones and physiological parameters in lactating cows.** J. Dairy Sci. 69 (Suppl. 1), 152.
Soderholm C.G., Otterby D.E., Linn J.G., Ehle F.R., Wheaton J.E., Hansen W.P. and Annexstad R.J. (1988) **Effects of recombinant bovine somatotropin on milk production, body composition and physiological parameters.** J. Dairy Sci. 71, 355-365.
Sonenberg M., Free C.A., Dellacha J.M., Bonadonna G., Haymowitz A. and Nadler, A.C. (1965) **The metabolic effects in man of bovine growth hormone digested with trypsin.** Metab. 14, 1189-1213.
Sonenberg M., Kikutani M., Free C.A., Nadler A.C. and Dellacha J.M. (1968) **Chemical and biological characterization of clinically active tryptic digests of bovine growth hormone.** Ann. NY Acad. Sci. 148, 532-558.
Sonenberg M., Yamasaki N., Kikutani M., Swislocki N.I., Levine L. and New M. (1972) **Studies on active fragments of bovine growth hormone.** In "Growth and Growth Hormone" (Eds. A. Pecile and E.E. Muller), Excerpta Medica, Amsterdam, pp. 75-90.
Spencer G.S.G. (1987) **Biotechnology in the potential practical application of somatotrophic hormones for improving animal performance.** Reprod. Nutr. Dévelop. 27 (2B), 581-589.
Struempler A.W. and Burroughs W. (1959) **Stilbestrol feeding and growth hormone stimulation in immature ruminants.** J. Anim. Sci. 18, 427-436.
Sullivan J.T., Taylor R.B., Huber J.T., Franson S.E., Hoffman R.G. and Hard D.L. (1988) **Relationship of production level and days postpartum to response of cows to sometribove, USAN (recombinant methionyl bovine somatotropin).** J. Dairy Sci. 71 (Suppl. 1), 207.
Sunshine P. (1977) **Digestion and absorption of proteins.** In "Mead Johnson Symposium on Perinatal and Developmental Medicine", n. 11, pp. 17-21.
Taylor A.M., Sharma A.K., Avasthy N., Duguid I.G.M., Blanchard D.S., Thomas P.K. and Dandona P. (1987) **Inhibition of somatomedin-like activity by serum from streptozotocin-diabetic rats : prevention by insulin treatment and correlation with skeletal growth.** Endocr. 121, 1360-1365.

Tessmann N.J., Kleinmans J., Dhiman T.R., Radloff H.D. and Satter L.D. (1988) **Effect of dietary forage : grain ratio on response of lactating dairy cows to recombinant bovine somatotropin.** J. Dairy Sci. 71 (Suppl. 1), 121.

Thomas C., Johnsson I.D., Fisher W.J., Bloomfield G.A. and Morant S.V. (1987a) **Effect of recombinant bovine somatotropin in milk production, reproduction and health of dairy cows.** Anim. Prod. 44, 460.

Thomas C., Johnsson I.D., Fisher W.J., Bloomfield G.A., Morant S.V. and Wilkinson J.M. (1987b) **Effect of somatotrophin on milk production, reproduction and health of dairy cows.** J. Dairy Sci. 70 (Suppl. 1), 175.

Thomas H., Green I.C., Wallis M. and Aston R. (1987) **Heterogeneity of growth-hormone receptors detected with monoclonal antibodies to human growth hormone.** Biochem. J. 243, 365-372.

Timoney J.F., Gillespie J.H., Scott F.W. and Barlogh J.E. (1988) **The virales.** In "Hagan and Bruner's Microbiology and Infectious Diseases of Domestic Animals", Cornell University Press, Ithaca, 8th edn, pp. 425-433.

Torkelson A.R., Dwyer K.A., Rogan G.J. and Ryan R.L. (1987) **Radioimmunoassay of somatotropin in milk from cows administered recombinant bovine somatotropin.** J. Dairy Sci. 70 (Suppl. 1), 146.

Torkelson A.R., Lanza G.M., Birmingham B.K., Vicini J.L., White T.C., Dwyer S.E., Madsen K.S. and Collier R.J. (1988) **Concentration of insulin-like growth factor I (IGF-I) in bovine milk. Effect of herd, stage of lactation and sometribove, USAN (recombinant methionyl bovine somatotropin).** J. Dairy Sci. 71 (Suppl. 1), 169.

Tsushima T. and Friesen H.G. (1973) **Radioreceptor assay for growth hormone.** J. Clin. Endocr. Metab. 37, 334-337.

Turman E.J. and Andrews F.M. (1955) **Some effects of purified anterior pituitary growth hormone on swine.** J. Anim. Sci. 14, 7-19.

Tyrrell H.F., Brown A.C., Reynolds P.J., Haaland G.L., Peel C.J., Bauman D.E. and Steinhour W.C. (1982) **Effect of growth hormone on utilization of energy by lactating Holstein cows.** In "Energy Metabolism of Farm Animals" (Eds. A. Ekern and F. Sundstol), EAAP, Publ. n 29, 46-47.

Tyrrell H.F., Brown A.C.G., Reynolds P.J., Haaland G.L., Bauman D.E., Peel C.J. and Steinhour W.D. (1988) **Effect of bovine somatotropin on metabolism of lactating dairy cows : energy and nitrogen utilization as determined by respiration calorimetry.** J. Nutr. 118, 1024-1030.

Van den Berg G. (1989) **Milk from BST-treated cows : its quality and suitability for processing.** In "Use of Somatotropin in Livestock Production" (Eds. K. Sejrsen, M. Vestergaard and A. Neimann-Srensen), Elsevier Applied Science, London and New York, pp. 178-191.

Van den Berg, G. and De Jong, E. (1986) **The influence of the treatment of lactating cows with methionyl bovine somatotropin on milk properties.** Report NOV-1209, Neth. Inst. Dairy Res., Ede.

Van der Hel W., Verstegen M.W.A., Huisman J., Kanis E., van Weerden E.J. and van der Wal P. (1988) **Effect of rpST treatment on energy balance traits and metabolic rate in pigs.** J. Anim. Sci. 66 (Suppl. 1), 225.

Van Es A.J.H. (1977) **The energetics of fat deposition during growth.** Nutr. Metab. 21, 88-104.

Vasilatos R. and Wangsness P.J. (1981) **Diurnal variations in plasma insulin and growth hormone associated with two stages of lactation in high producing dairy cows.** Endocr. 108, 300-304.

Vassal L. (1988) **Fabrication de fromages de type Camembert et St. Paulin à partir de lait de vaches traitées (E) ou non (T) à la somatotropine, resumé des principales observations.** Rapport préliminair, INRA, Stat. Rech. Lait., Jouy-en-Josas.

Verde L.S. and Trenkle A. (1987) **Concentrations of hormones in plasma from cattle with different growth potentials.** J. Anim. Sci. 64, 426-432.

Vernon R.G. (1982) **Effects of growth hormone on fatty acid synthesis in sheep adipose tissue.** Int. J. Biochem. 14, 255-258.

Vernon R.G. (1986). **The response of tissues to hormones and the partition of nutrients during lactation.** Hannah Res. Inst., 115-121.

Vernon R.G. (1988) **The partition of nutrients during the lactation cycle.** In "Nutrition and Lactation in the Dairy Cow" (Ed. P.C. Garnsworthy), Butterworth, London, pp. 32-52.

Vernon R.G. (1989) **Influence of somatotropin on metabolism.** In "Use of Somatotropin in Livestock Production" (Eds. K. Sejrsen M. Vestergaard and A. Neimann-Srensen), Elsevier Applied Science, London and New York, pp. 31-50.

Vernon R.G. and Finley E. (1985) **Regulation of lipolysis during pregnancy and lactation in sheep.** Biochem. J. 230, 651-656.

Vernon R.G. and Finley E. (1988) **Roles of insulin and growth hormone in the adaptations of fatty acid synthesis in white adipose tissue during the lactation cycle in sheep.** Biochem. J. 256, 873-878.

Vernon R.G. and Flint D.J. (1989) **Role of growth hormone in the regulation of adipocyte growth and function.** In "Biotechnology in Growth Regulation" (Eds. R.B. Heap, C.G. Prosser and G.E. Lamming), Butterworth, London, pp. 57-71.

Vernon R.G. and Taylor E. (1988) **Insulin, dexamethasone and their interactions in the control of glucose metabolism in adipose tissue from lactating and non-lactating sheep.** Biochem. J. 256, 509-514.

Vernon R.G. Clegg R.A. and Flint D.J. (1981) **Metabolism of sheep adipose tissue during pregnancy and lactation.** Biochem. J. 200, 307-314.

Vernon R.G., Faulkner A., Finley E., Pollock H. and Taylor E. (1987a) **Enzymes of glucose and fatty acid metabolism of liver, kidney, skeletal muscle, adipose tissue and mammary gland of lactation and non-lactating sheep.** J. Anim. Sci. 64, 1395-1411.

Vernon R.G., Finley E. and Flint D.J. (1987b) **Role of growth hormone in the adaptations of lypolysis in rat adipocytes during recovery from lactation.** Biochem. J. 242, 931-934.

Vezinhet A. (1973) **Effect of hypophysectomy and bovine somatotrophic hormone treatment on the relative growth of lambs.** Ann. Biol. Anim. Bioch. Biophys. 13, 51-73.

Vicini J.L., Clark J.H., Hurley W.L. and Bahr J.M. (1988) **Effects of abomasal or intravenous administration of arginine on milk production, milk composition, and concentrations of somatotropin and insulin in plasma of dairy cows.** J. Dairy Sci. 71, 658-665.

Vignon C.S. (1987) **Injection of somatotropin to grazing dairy cows in 1986. Effect on the composition of fats and nitrogenous substances in milk.** Concluding experimental report, Inst. Nat. Polytech. Lorraine, Nancy.

Vignon C.S. (1988) **Effect of the injection of somatotropin in dairy cows on the composition and technological value of milk.** End of experiment report, Inst. Nat. Polytech. Lorraine, Nancy.

Vignon C.S. and Ramet J.P. (1988) **Effect on the technological value of milk by injecting dairy cows with somatotropin.** End of experiment report, Inst. Nat. Polytech. Lorraine, Nancy.

Vukavic T. (1984) **Timing of the gut closure.** J. Pediatr. Gastroenterol. Nutr. 3, 700-703.

Wagner J.F. and Veenhuizen E.L. (1978) **Growth performance, carcass deposition and plasma hormone levels in wether lambs treated with growth hormone and thyroprotein.** J. Anim. Sci. 45 (Suppl. 1), 397.

Wagner J.F., Cain T., Anderson D.B., Johnson P. and Mowrey D. (1988) **Effect of growth hormone (GH) and estradiol ($E_2B$) alone and in combination on beef steer growth performance, carcass and plasma constituents.** J. Anim. Sci. 66 (Suppl. 1), 283-284.

Wallace A.L.C. and Bassett J.M. (1966) **Effect of sheep growth hormone on plasma insulin concentration in sheep.** Clin. Expt. Metab. 15, 95-97.

Wallis M. (1989) **Species specificity and structure-function relationships of growth hormone.** In "Biotechnology in Growth Regulation" (Eds. R.B. Heap, C.G. Prosser and G.E. Lamming), Butterworth, London.

Walton P.E. and Etherton, T.D. (1986) **Stimulation of lipogenesis by insulin in swine adipose tissue : antagonism by porcine growth hormone.** J. Anim. Sci. 62, 1584-1595.

Walton P.E., Etherton T.D. and Evock C.M. (1986) **Antagonism of insulin action in cultured pig adipose tissue by pituitary and recombinant porcine growth hormone : potentiation by hydrocortisone.** Endocr. 118, 2577-2581.

Wardzala L.J., Simpson I.A., Rechler M.M. and Cushman S.W. (1984) **Potential mechanism of the stimulatory action of insulin on insulin-like growth factor II binding to the isolated rat adipose cell.** J. Biol. Chem. 259, 8378-8383.

Warriss P.D. and Kestin S.C. (1988) **Beta-agonists improve the carcass but may reduce meat quality in sheep.** Anim. Prod. 46, 502.

Watt P.W., Clegg R.A., Flint D.J. and Vernon R.G. (1989) **Increases in sheep b-receptor number on exposure to growth hormone in vitro.** In "Biotechnology in Growth Regulation" (Eds. R.B. Heap, C.G. Prosser and G.E. Lamming), Butterworth, London.

Weber T. (1988) **Wirkung von rekombinantem bovinen Somatotropin (BST) bei Milchkühen in zwei aufeinanderfolgenden Laktationen.** Diss. Kiel, 152 p.

Webster A.J.F. (1985) **Farm Animal Welfare. The needs of animals and the wishes of society.** British Society of Animal Production : Winter meeting, Paper No. 39.

Weetman A.P. and McGregor A.M. (1984) Autoimmune thyroid disease : **developments in our understanding.** Endocr. Rev. 5, 309-355.

West E.S., Todd W.R., Mason H.S. and Van Bruggen J.T. (1966) "Textbook of Biochemistry", The Macmillan Co., New York.

West J.W., Johnson Jr. J.C. and Bondari K. (1988) **The effect of bovine somatotropin on productivity and physiologic responses of lactating Holstein and Jersey cows.** J. Dairy Sci. 71 (Suppl. 1), 209.

Wheatley I.S., Wallace A.L.C. and Bassett J.M. (1966) **Metabolic effects of ovine growth hormone in sheep.** J. Endocr. 35, 341-353.

Wheaton J.E., Al-Raheem S.N., Massri Y.G. and Marcek J.M. (1986) **Twenty-four-hour growth hormone profiles in angus steers.** J. Anim. Sci. 62, 1267-1272.

Whitaker D.A., Smith E.J., Kelly J.M. and Hodgson-Jones L.S. (1988) **Health, welfare and fertility implications of the use of bovine somatotrophin in dairy cattle.** Vet. Rec. 122, 503-505.

White T.C., Lanza G.M., Dyer S.E., Hudson S., Franson S.E., Hintz R.L., Duque J.A., Bussen S.C., Leak R.K. and Metzger L.E. (1988) **Response of lactating dairy cows to intramuscular or subcutaneous injection of sometribove, USAN (recombinant methionyl bovine somatotropin) in a 14-day prolonged release system. Part I. Animal performance and health.** J. Dairy Sci. 71 (Suppl. 1), 167.

Catecholamine, 11, 16, 71
Cattle, 6, 16, 18-19, 24, 26-27, 30, 54, 67, 75
Cell, 2, 9-10, 13, 18, 22, 29, 39-40, 58, 65, 67-70
Cellulose, 7, 22
Chain, 1, 3, 36, 42, 52, 54, 56, 61, 74
Charge, 2
Cheese, 41-42, 44, 51
Chymotrypsin core, 60-61
Cimaterol, 69
Citrulline, 7
Clinical disease, 29-30, 53, 72
Clinical examination, 27
Clonidine, 65
$CO_2$, 28
Coagulation, 44
Cobalt, 24
Collagen solubility, 55-56, 66, 76
Colostrum, 54, 66
Concentrate, 23, 31, 54
Consumer, 37-38, 45, 52, 54, 63
Control, 2, 19, 32-33, 43, 48, 68, 74
Controlling milk quality, 30, 73
Cooking, 52, 55-56, 76
Copper, 42
Cow, 3-5, 7-9, 11, 14-16, 18-25, 27-38, 41-45, 47, 49-57, 62, 65-68, 71-74
Cow breed, 33, 35, 48-50, 54-55
Cow health, 19, 21, 27, 29, 32, 51, 72
Cow reproduction, 19, 30, 32, 72
Cow welfare, 27, 32, 52, 72
Cu, 73
Cysteine, 42

## D

D-galactose, 39-41
D-galactose 1-phosphate, 40-41
D-glucose, 6, 7, 39-41
D-lactose, 39
Dairy cattle, 19, 30
Dairy cow, 1, 9, 11, 19, 21, 24, 27-29, 35, 42, 44, 52-54, 56, 65-66, 72
Dairy industry, 30
Days open, 31, 67
Deoxy-adenosyl-cobalamine, 28
Development, 10, 28, 32, 53, 72
Diabetes, 43
Diet, 14, 22-23, 30, 54, 64
Dietary protein, 22
Digestive treat, 14, 61
Disaccharide, 39
Disease, 30, 32-34, 41, 52-53, 64, 72, 73
DNA, 1, 66, 70
Dosage, 19-20, 29, 31, 42, 49-51, 57, 75
Dwarf, 58
Dwarfism, 43

## E

Early lactation, 14, 20-22, 66-67, 72
Endocrine system, 1, 2, 11, 15, 66
Endopeptidase, 42, 73
Energy, 3, 5-6, 8, 15, 19-24, 35, 72
Energy balance, 5-6, 8, 14, 16, 18-22, 24, 30, 33-34, 38, 41, 53, 71-72
Energy state, 4, 15
Enkephalins, 65
Environmental condition, 24, 30, 32-34, 64, 66
Enzyme, 6-7, 28, 38-42, 60-61
Enzyme activity, 42, 73
Eumetabolic action, 14, 17-18, 22, 24
Eumetabolic effect, 16
Exploitation, 33
Extra-dietary nutrients, 18, 71
Extra-milk, 5, 19

## F

Facilities, 34, 72
Farm, 28, 30, 32-37, 50
Farmer, 32, 51-53, 63
Fat, 5-6, 16, 20, 22, 24, 35, 38, 41-42, 45, 53-55, 63, 68, 70-71, 73
Fat deposition, 68, 69
Fat mobilization, 41
Fatty acid, 5-8, 13, 15-18, 20, 28-29, 38, 41, 54, 71
Fe, 73
Feed, 4, 14, 19, 21-22, 24-25, 64, 70-71
Feed intake, 5, 20-22, 24, 28, 53
Feed stuff, 69
Feedback, 2, 70
Fermentation, 7, 38, 54
Fertility, 35, 72
Fetus, 32
Fibroblasts, 4
Fluid filled bursal swelling, 27, 35, 53
Food, 28, 30, 36, 37, 47, 52, 55, 61, 63-64, 68
Food intake, 14-16, 18, 38, 71
Forage, 21-22, 25, 54
Fragment, 51, 60-62, 65, 75
Free energy, 6
Freezing point, 43
FSH, 2-3
Fumarate, 29
Fusion genes, 70

## G

Galactokinase, 40
Galactopoietic effect, 4, 9, 71

Galactopoietic properties, 1
Galactosemia, 39, 41
Galactosyl-transferase, 6, 36, 39
Gastrointestinal tract, 51, 60-61
Gene, 70
Genetic selection, 34-35, 44, 72-73
Genome, 70
Gigantism, 43
Gland epithelium, 10
Glucagon, 3, 15, 69
Gluconeogenesis, 5-7, 16
Glucose, 5-18, 13, 15-18, 28, 38, 68
Glucose oxidation, 13, 15, 17-18, 71
Glutamic acid, 59
Glycine, 59
Glycolytic enzyme, 6
Grass silage, 21, 23
GRF, 64-65
Growing animals, 14-15, 50, 73-74
Growth, 14, 28, 43, 47-50, 52-53, 60, 72
Growth hormone, 60, 62, 64, 68-70, 75
GTP, 6, 29
Guidelines, 33, 72
Gut, 57, 61-62, 75

## H

Half-life, 56-57, 68, 74
Handling, 34, 44, 69
Hay, 21
Health, 19, 21, 27, 29-30, 32, 51, 72, 74
Heart rate, 27, 32
Heifers, 15, 23, 25, 48-49, 55, 71
Hematological alteration, 29-30
Hexapeptide, 65
High-yielding animals, 31
High-yielding cow, 18, 31, 34-35, 72
Histidine, 59
Homeorhetic action, 14
Homeorhetic effect, 16
Hormone, 1-4, 11, 13, 15, 39, 42-43, 47, 50, 59-60, 62, 64-71, 75
hST, 58
Human breast milk, 50-51, 74
Human gut, 57, 62, 75
Human subjects, 51, 53, 60-62
Humans, 3, 24, 36, 39-40, 43, 45, 50-53, 58-62, 64, 69, 74-75
Hypopituitary humans, 75, 62
Hypothalamic hormone, 1-2
Hypothalamus, 1-2

## I

IGF-I, 3-4, 9-11, 14-16, 18, 39, 43, 47-52, 65-68, 71, 74

IGF-I in milk, 9-10, 50-51, 66-67, 74
IGF-II, 11, 43, 51-52, 66-67
Immunization, 53, 69
Immunogenic effects, 52-53
Immunoglobulin, 53, 69
Immunosuppression, 52
Injection, 9, 19-20, 24, 27, 32-34, 41, 43, 49, 53, 57, 62, 70, 72-74
Injection sites, 33, 56-57, 72
Injection procedure, 34
Insulin, 4, 11, 15-16, 29, 43, 65-68, 71
Insulin-like growth factor, 3, 9, 43, 45, 48, 51-52, 66-67, 71, 74
Insulin-resistance, 18
Intestinal tract, 40, 51, 74-75
Iron, 42
Iso-leucine, 59
Isoelectric point, 58

## J

Juice extraction, 55-56, 76
Juiciness, 55

## K

Keto-acidosis, 29
Ketosis, 28-29, 35, 72-73
Kidney, 6

## L

Labelling, 56, 74
Lactase, 39-40
Lactate, 6-7, 17
Lactating animals, 3, 5, 8, 14-15, 17-18, 64-65, 71
Lactating cattle, 67
Lactating cows, 1, 3, 5, 9, 11, 14-15, 18, 27, 35, 57, 68, 72
Lactation, 8-9, 11, 14, 16, 20-24, 27, 29, 31, 33, 35, 39, 47-49, 53-54, 66-67, 72
Lactation yield, 31, 71
Lactogenic receptor, 4
Lactose, 6-7, 20, 15, 28, 38-41
Lactose intolerance, 40
Lactose synthase, 39
Lemeness, 27, 34, 73
Lean, 16, 55, 70
Lean tissue, 69
Leu-enkephalin, 65
Leucine, 59, 65

107

LH, 2-3
Lipase, 42, 73
Lipid, 5, 14, 16-18, 20, 22, 71
Lipid mobilization, 18, 20, 22, 53
Lipid metabolism, 5-6, 13-14, 16-17
Lipogenesis, 14, 16, 18
Lipolysis, 6, 11, 14, 16
Lipolytic effect, 11, 16, 53, 67-68
Lipoprotein lipase, 16
Liver, 4, 6-7, 11, 16, 40-41, 69-70
Long-term effects, 19, 24, 68
Long-term treatment, 16, 28, 30, 49, 55
Lymphocytes, 4
Lysine, 23, 59

## M

Magnesium, 42
Mammary blood flow, 8-11, 17
Mammary epithelium, 5, 10
Mammary gland, 4, 6, 8-11, 14-15, 17, 35, 38-39, 41-42, 64, 67, 71
Mammary secretion, 66-67
Mammary tissue, 4-5, 9, 11, 18, 38, 66-67, 71
Management, 4, 24-25, 28, 30, 33-34, 52, 72
Management conditions, 23, 29, 71-72
Manipulation, 63-70
Mass, 2, 70
Mastitis, 30, 34-35, 52, 72
Meat, 30, 47, 55-57, 60, 62-63, 74-75
Meat quality, 36, 55-56, 63, 69, 75
Met-enkephalin, 65
Metabolic adaptation, 15
Metabolism, 5-6, 8, 11, 13-14, 16-19, 22, 28-29, 42-43, 57, 66, 68, 72
Metabolization, 57
Methionine, 7, 11, 23, 59, 65
Methionyl-BST, 11
Mg, 40, 42, 73
Micelles, 42
Microorganism, 7, 38
Mid-lactation, 8, 20, 31, 66
Milk, 5-7, 9, 20-21, 23, 30, 35, 37-38, 41-45, 47-56, 60, 62-67, 70-74
Milk fat, 5-6, 20, 22, 24, 38, 41
Milk fever, 28-29
Milk processing, 42, 44-45
Milk production, 1, 4-5, 9-11, 14-25, 28-35, 38, 42, 44, 48
Milk quality, 30, 36, 41, 43, 45
Milk secretion, 1, 9-10, 23, 25
Milk synthesis, 5, 17-18
Milk yield, 5, 8-11, 14, 17, 20, 23-24, 28, 32, 34-35, 44
Mineral, 13, 20
Misuse, 33-35, 72
Mitochondria, 7, 16

Mn, 73
Mobilization, 11, 18, 20-22, 41, 53-54, 71
Modulatory action, 8, 18
Molecular weight, 10, 58, 60, 66, 68
Monkey, 58-59
Monoclonal antibodies, 68
Multiparous cow, 25, 71
Muscle, 5, 14-15, 18, 21, 55-56, 69-70, 74

## N

N-acetyl-glucosamine, 38
Na, 42, 73
Native IGF-I, 66
Natural characteristics, 64
NEFA, 6, 8
Negative energy balance, 5-6, 8, 14, 16, 18, 21, 24, 30-31, 33-34, 38, 41, 53, 71-72
Neuroendocrine system, 3, 43
Non-esterified fatty acids, 6, 8, 15
Non-ruminants, 4
Nucleic acid, 13
Nutrient, 4, 8, 10-11, 14, 17-18, 20-21, 23, 25, 31, 66, 68, 71-72
Nutrient partitioning, 71
Nutrition, 19, 24, 29-30, 32-33, 72

## O

Oral ingestion, 61-62, 75
Organizational problem, 33
Organoleptical characteristic, 37-38, 43-45, 56
Ornithine, 7
Overall acceptability, 55-56, 76
Overstretched cull cows, 33, 72
Oxaloacetate, 6, 28-29
Oxidation, 6-7, 13, 15-18, 28, 71

## P

P, 42, 72-73
Pain, 33, 65
Painful, 72
Pancreas, 68
Pancreatic hormone, 68
Pancreatic islets, 4
Pasteurization, 51-52, 74
Peak-lactation, 3, 8-9, 22, 49
Periods of lactation, 17, 19, 30-31, 44
Peripheral tissue, 4, 6, 10, 15, 18, 71
Persistency, 20, 23, 35, 49, 73
Pesticides, 53

pH, 22, 38, 43, 55, 58
Phase of lactation, 29
Phenyl-alanine, 58-59
Phosphatase, 42, 73
Phosphoenolpyruvate, 6
Phosphoglucomutase, 41
Phosphorus, 28, 42
Pigmentation, 55
Pituitary, 1-3, 11, 43, 56, 58, 62, 67, 69, 74
Pituitary somatotropin, 1, 3, 58-59, 62, 66, 75
Plane of nutrition, 72
Plasma, 3, 6, 8-10, 15-16, 30, 47-49, 51, 56-57, 67
Plasma level, 3, 15-16, 48, 50-51, 73-74
Plasma volume, 29, 72
Polypeptides, 1-2, 68
Positive energy balance, 5-6, 14
Post-partum, 35, 49-50, 54, 66, 72
PPi, 28, 41
Precursor, 7, 17, 28, 39, 50, 71
Pregnancy, 23, 31, 66-67
Processing, 37-38, 41, 44-45, 47, 51-52, 54, 73-74
Processing suitability, 38, 73
Prolactin, 3-4, 15, 65
Prolonged-release BST, 38, 49, 56, 73
Prolonged-release preparation, 20-21, 24, 29, 49-50, 57
Propionate, 6-7, 28
Propionyl-CoA, 28
Protease, 42, 73
Protein, 1, 3, 5, 10, 13-14, 20-23, 35, 41-42, 52-53, 56, 61, 66, 68, 73-74
Protein mobilization, 21
Protein synthesis, 13, 69
Pyridoxine, 43
Pyruvate, 6, 15, 41, 73

### Q

Quality control, 30

### R

r-RNA, 5
Ractopamine, 69
Radio-immunoassay, 48
Rat, 4, 59-61, 70
Receptor, 2-4, 9, 16, 18, 51, 59, 65-69
Receptor binding, 3-4
Red cells, 29, 72
Renneting reaction, 44
Reproduction, 19, 30, 32, 72
Reproductive efficiency, 30, 72
Reproductive performance, 21, 25, 27, 30-32
Requirement, 4, 20-24, 33, 63, 71
Residual concentration, 74

Residue, 2, 39, 41-42, 53, 56, 58
Resolution time, 33
Respiratory rate, 27, 33
Retrovirus, 70
RIA, 48
Riboflavin, 43
Routine control, 32
Rumen, 7, 22-23, 54
Ruminants, 1, 4, 6-7

### S

Safety, 52
Season, 22, 54, 67
Seasonal changes, 54
Selection, 34-35, 44, 73
Semantic aspect, 64
Sequence, 3, 11, 40-41, 51, 59, 62, 65, 70, 75
Serine, 59
Serum, 42, 68
Shear force, 55, 76
Skim milk, 48, 50, 51, 73
Small intestine, 39
Sodium, 42
Somatic cell count, 30, 43-44, 72-73
Somatogenic receptor, 4
Somatomedin, 9, 11, 16, 39, 43, 45, 47, 50-51, 66-67
Somatotropin, 1-5, 9, 11, 14, 17-19, 22, 24, 28-29, 31, 34-35, 43, 45, 47-50, 53, 55-60, 62, 64-70, 73-75
Somatotropin fragment, 62, 65
Sparing action, 5-6, 8, 11, 18, 71
Stability, 38, 44
Starch, 22
Steer, 48, 55-56
Sterilization, 51-52, 74
Stockmen, 34
Stressing effect, 52
Subclinical condition, 29-30
Succinate, 7, 29
Super-mouse, 70
Swelling, 33, 72
Syneresis, 44

### T

Tagging, 56, 74
Taste, 39, 55, 76
Tertiary structure, 3, 42
Thiamine, 43
Threonine, 59
Thyroid gland, 2
Thyrosine, 59
Thyrotropin, 2
Thyroxine, 2-3, 15
Tissue swelling, 72

Toxin, 69
Transcription, 70
Transformed embryonic stem, 70
Transgenic approach, 70
Transgenic animal, 70
TRH, 2
Triacylglycerols, 28
Triiodothyronine, 2-3, 15
Truncated form of IGF-I, 66-68
Tryptic digest, 60
Tryptic digestion, 60
Tryptophane, 59
TSH, 2-3, 69

## U

UDP, 39
UDP-D-galactose, 39-41
UDP-D-glucose, 40, 41
UDP-glucose 4-epimerase, 41
UDP-glucose pyrophosphatase, 41
UDP-glucose:α-D-galactose 1-phosphate, 40, 41
Undernourishment, 33
Undiagnosed disease, 32, 72
Urea, 7, 8, 44

## V

Valine, 7, 59
Vascular system, 11
Veterinarian, 24
Veterinary control, 32-33
Veterinary supervision, 32-33, 72
Virus, 52
Vitamin A, 43
Vitamin $B_{12}$, 24, 28, 43
Vitamin, 42-43, 64, 73
Volume of plasma, 29, 72

## W

Weight gain, 68-69
Welfare, 20, 24, 27, 32, 52, 70, 72
Whey protein, 42, 53, 73

## Z

Zinc, 42
Zn, 73

Photocomposition et impression
IMPRIMERIE LOUIS-JEAN
BP 87 — 05003 GAP Cedex
Tél. : 92.51.35.23
Dépôt légal : 580 — Septembre 1990
Imprimé en France